SUSTAINABLE DEVELOPMENT

Principles, Frameworks, and Case Studies

Social-Environmental Sustainability Series

Series Editor

Chris Maser

Published Titles

Social-Environmental Planning: The Design Interface Between Everyforest and Everycity, **Chris Maser**

Sustainable Development: Principles, Frameworks, and Case Studies, **Okechukwu Ukaga, Chris Maser, and Michael Reichenbach**

SUSTAINABLE DEVELOPMENT

Principles, Frameworks, and Case Studies

Edited by
Okechukwu Ukaga, Chris Maser,
and Mike Reichenbach

CRC Press
Taylor & Francis Group
Boca Raton London New York

CRC Press is an imprint of the
Taylor & Francis Group, an **informa** business

CRC Press
Taylor & Francis Group
6000 Broken Sound Parkway NW, Suite 300
Boca Raton, FL 33487-2742

First issued in paperback 2019

ISBN-13: 978-1-4398-2062-9 (hbk)
ISBN-13: 978-0-367-38454-8 (pbk)

Library of Congress Cataloging-in-Publication Data

Sustainable development : principles, frameworks, and case studies / edited by Okechukwu Ukaga, Chris Maser, Mike Reichenbach.
p. cm. -- (Social environmental sustainability)
Includes bibliographical references and index.
ISBN 978-1-4398-2062-9
1. Sustainable development. 2. Environmental economics. 3. Environmental policy--Social aspects. I. Ukaga, Okechukwu, 1961- II. Maser, Chris. III. Reichenbach, Mike. IV. Title. V. Series.

HC79.E5S86467 2010
338.9'27--dc22 2009030998

Visit the Taylor & Francis Web site at
http://www.taylorandfrancis.com

and the CRC Press Web site at
http://www.crcpress.com

Contents

Series Editor's Note

We (Okechukwu Ukaga, Chris Maser, and Michael Reichenbach) have come together out of a sense of trusteeship for the generations of children of today, tomorrow, and beyond to create a book that will help protect for them a future of good quality.

The purpose of our book is to frame in the positive some of the most important, human-related aspects of sustainable development. There is great power in learning to think in a positive mode. In so doing, members of a community can create a shared vision whereby they not only understand their community from several vantage points, but also understand that much of the confusion in communication comes from trying to move away from negatives. Trying to move away from a negative precludes people from saying what they really mean because they are focused on what they do *not* want. As long as people express what they do not want, it is virtually impossible to figure out what they *do* want because they are "symptomatic" in their thinking as opposed to "systemic."

In other words, we continually focus on and attempt to treat the symptoms of today's emerging global crises, but we refuse to deal with the cause—which in part is the opulence of our Western industrial lifestyle. These crises include a warming climate, per-capita shrinking of natural resources due to a human population that is rapidly exceeding the global carrying capacity, and the ever-growing disparity between the rich and poor peoples and nations. We cannot, however, move away from an unwanted, negative circumstance. We can only move toward a desired, positive outcome.

To help us elucidate our common journey toward a positive future for all generations, we have invited distinguished practitioners and scholars to contribute their expertise to help reveal the multidimensional nature of sustainable development. Their approaches include strong theoretical and historical pieces as well as salient case examples, which illustrate the outcome of each model or framework.

Chris Maser
Series Editor

Preface

Okechukwu Ukaga, Chris Maser, and Michael Reichenbach

An environmental disaster may be looming within many of our readers' lifetimes and certainly within the lifetime of the next generation. The full impacts of Earth climate change are not known, nor can they be known. Carbon emissions are linked to Earth warming, the impacts of which include rising oceans, flooded coastlines, changes to where and how we grow our crops, and changes in the ecology of the planet. As we write this in the summer of 2009, the Earth economic system is in the process of correction. While we will, as in the past, recover from this latest downturn, it is yet another powerful reminder that economic growth cannot continue indefinitely. To be sustainable, all societies must adjust to new realities, which include changing ecosystems and natural limits to growth. How do we address these issues and maintain an equitable way of life for all on the planet? The solutions are embedded in the principles of sustainable development. Hence, this book presents frameworks and cases that have been used or could be used to address impacts of climate change, challenging economic conditions, social problems, and other complex issues related to our future.

While the basic idea behind sustainable development is not new, especially among traditional or indigenous societies, the present-day concept of sustainability can be traced to the United Nations Conference on Human Environment, held June 5–16, 1972, in Stockholm, which highlighted the link between the environment and development in the Earth arena (United Nations General Assembly 1972, 148). The World Commission on Environment and Development (1987, 383) further catalyzed the Earth conversation about sustainable development, which it defined in their publication *Our Common Future* as "development that meets the needs of the present without compromising the ability of future generations to meet their own needs." This recognizes the complex interrelationships among the environment, people, and development, as well as the past, present, and future generations.

Sustainability, itself a transcendent term, is an ideal toward which we strive and not an endpoint. It has many interrelated dimensions (e.g., ecological, economic, social, political, and epistemological) and calls for a participatory, holistic, interdisciplinary approach to program planning, implementation, and evaluation. Sustainable development requires the freedom to openly discuss, challenge, and debate ideas and perceptions as points of greater learning, as opposed to blind defenses of entrenched positions. Sustainable development also requires the ability to think about systems and their interaction. Sustainable development is created anew in the eternal moment of every day, and thus is an ideal toward which we strive that cannot be realized as a given point in time. Working toward this vision is a challenging task—one that requires not only serious effort, but also appropriate tools and systems for moving effectively from improved awareness, interest, knowledge, attitude, and skills to practical steps toward sustainability. In other words, those interested in fostering sustainable development will benefit greatly from concrete ideas of what can be done on the ground to meet critical needs.

There is an African proverb that states: It takes a village to raise a child. This maxim holds a truth that extends well beyond the village. Namely, for a village to be healthy, it requires a unifying and holistic factor to integrate the myriad interactive components into a functional whole. And that unifying factor can be the concept of sustainable development.

If we were to ask ourselves what it would take to design a sustainable community, country, or world, we would find ourselves engaging the whole system (instead of a few parts) over the long term (instead of the short run). Moreover, for life, human or otherwise, to have any measure of good quality, the basic components of long-term sustainability of the Earth commons must be given the highest priority: from clear air, to pure water, healthy oceans, fertile soils, and healthy food, to asking the children what kind of future they want their parents—as trustees—to protect for them as a legacy from one generation to the next.

What, you might ask, is meant by the "Earth commons?" The commons is that part of the world that is every person's birthright. As such, there are two kinds of commons. Some are gifts of nature, such as clean air, pure water, fertile soil, a rainbow, a beautiful sunset, or a tree growing in the middle of a village; others are the collective product of human creativity, such as the town well from which everyone draws water. The commons provides the basic ecological and social support systems of life and well-being. It's the vast realm of our shared heritage, which we typically use free of toll or price. Air, water, and soil; sunlight and warmth; wind and stars; mountains and oceans; languages and cultures; knowledge and wisdom; peace and quiet; sharing and community; joy and sorrow; and the genetic building blocks of life—these are all aspects of the commons.

A commons has an intrinsic quality of just being there, without formal rules of conduct. People are free to breathe the air, drink the water, and share life's experiences without a contract, without paying a royalty, without needing to ask permission. As such, a commons engages people in the wholeness of themselves and in community. It fosters the most genuine of human emotions and stimulates interpersonal relationships in order to share the experience, which enhances its enjoyment and archives its memory. We humans have jointly inherited the commons, which is more basic to the quality of our lives than either the market or the state. We are "temporary possessors and life renters," wrote British economist and philosopher Edmund Burke, and we "should not think it amongst [our] rights to cut off the entail, or commit waste on the inheritance."[1]

The foregoing leads us to the problem we face today, which is the myth of limitless resource. It is time we shift our defiance of limits to an embrace of "enoughness." All nations—but especially the wealthy, industrialized nations—must now consciously accept and openly display their social responsibility by placing personal and national limits on material appetites, which means separating need from want, and thereby learning when enough is, in fact, enough. As such, *limits* and *enoughness* are not only the bedrock of sustainability, but also its two most widely ignored components. There is, after all, no such thing as absolute personal or national freedom in anything. All choices have consequences for which we are responsible, a condition from which no one can escape.

By analogy, consider that you are grossly overweight and your doctor tells you to lose fifty pounds or risk the certainty of a fatal heart attack. You think about what the doctor said and decide that you must go on a strict diet, which you do. As a result, you lose ten pounds in a week, and you are amazed at how easy it is. The next five pounds requires two weeks to shed, and another ten requires a month. Although thrilled at losing twenty pounds, which makes you feel much better about yourself, you refuse to give up your old, comfortable, sedentary lifestyle. Then, much to your dismay, the weight begins once again to accumulate, until you are back where you started.

Clearly, the initial weight loss was but a temporary fix of a symptom, which is why you went to your doctor in the first place. However, as the doctor admonished, you have to fundamentally change your lifestyle not only to reduce all your excess weight, but also, and more importantly, to keep it off. That, however, was *not* what you wanted to hear, so you resisted that part of the prescription. Finally, with your health permanently at stake, you go back to your doctor and, together, create a plan that will allow you to alter your lifestyle in relatively nonthreatening steps over a given period of time. And that is what we are prescribing in this book—a plan that begins with an appropriate assessment and understanding of prevailing conditions, learning from the many examples of sustainable development processes and practices that can be applied to your own situation, and personal decisions and leadership that gradually extend to a community, then a region, a state, a nation, and ultimately to the world. Here, it must be understood that it is necessary to accept every step toward sustainability, however small, as a vital, positive contribution to the whole.

We, the peoples of the Earth, must find ways to work together to raise the level of our collective consciousness to sustain the dignity of being human. Sustainable development is not about maintaining a good quality of life for just one generation, but rather about passing the ability to realize a good quality of life from generation to generation. Our quality of life is a construct of our existence. Each generation—as a trustee for the next generation—is charged with passing forward the intact principle of the trust. Increasing the quality of life is not a promise of sustainable development. While we may desire that our quality of life increase, we need to also be committed to ensuring equity in meeting all peoples' basic requirements—present and future.

We, therefore, begin the text with discussions about valuation. The concept of valuation is important since people need to understand the context in which they are working in order to figure out what they need to change or sustain and how. In his contribution entitled "If the GDP Is Up, Why Is the GPI Down?" Venetoulis highlights the disconnect between gross domestic product (as a measure of economic well-being) and true social-environmental welfare. The chapter demonstrates how genuine progress indicators (GPIs) respond to the theoretical and practical defects inherent in the gross domestic product as a measure of economic sustainability, making genuine progress indicators a more honest way to inform public policy with respect to such things as globalization, tax cuts, and urban sprawl.

In Chapter 2, Talberth and Cobb present the conceptual framework, methods, and results for the 2006 update of the genuine progress indicator (GPI) for the United States. The genuine progress indicator is one of the first alternatives to gross domestic product vetted by the scientific community and used regularly by government and nongovernmental organizations worldwide. The GPI's structure is grounded in the principles of sustainable development, which call for (1) no net loss of natural capital, (2) welfare-based accounting, (3) distributional equity, and (4) throughput minimization. The genuine progress indicator relies on the same personal consumption data as the gross domestic product, but makes deductions to account for costs associated with negative factors, such as income inequality, crime, environmental degradation, loss of leisure, and international borrowing; and additions reflecting the value of services from consumer durables, public infrastructure, volunteering, housework, and ecological restoration. The genuine progress indicator suggests that, while the scale of U.S. economic activity has grown steadily since 1950, our collective welfare may have peaked in the late 1970s and stagnated ever since, as the benefits of economic growth since that time have been offset by increasing income inequality, costs associated with climate change, depletion of forests and farmland, and a worsening international position.

Chapter 3, by Venetoulis and Talberth, clarifies the definition of ecological footprint applications and proposes several methodological and theoretical refinements. Their new approach includes the biological capacity of the entire surface of the Earth, allocates space for other species, changes the basis of equivalence factors to net primary productivity (NPP), reallocates the carbon budget, and reports biological capacity of carbon sequestration. The ecological footprint measures how much of the biosphere's annual regenerative capacity is required to renew the natural resources used by a defined population in a given year.

In Chapter 4, Bawden and Reichenbach present systemic and futures thinking as a way to transform our worldviews. The chapter is concerned with social learning and how viewpoints held by individuals influence both individual and group actions toward sustainability.

Critical scenario learning for systemic development uses scenario learning as a tool to move groups toward sustainable development. It is the antithesis of the typical approach to planning, where effort is placed on trying to create a desired future condition. The chapter asserts that we do not always have the control needed to create that desired future condition; thus, we need to envision a variety of futures, create a rich and rigorous prehistory for each, and use what is learned to test our current strategies. This framework utilizes a social process of reflexive, experiential, and transformational learning. For the framework to be used, participants must be able to recognize other viewpoints and allow those viewpoints to be challenged. The underlying assumptions for the framework are described. A case is provided illustrating how participants in the process learn their way toward sustainable development (i.e., the restoration of coniferous forests along the North Shore of Lake Superior).

We can use our unique ability as human beings to perceive multiple futures, examine them, explore viewpoints, and learn from the experiences. Future learning requires one to be mindful of systemic connections between ourselves, as humans, and the natural and social environments in which we are embedded. It is a mentally challenging process that leads to personal development as the essential prerequisite for sustainable social and material development.

As a central theme of this book, sustainable development is about raising our collective level of consciousness to the point where we can move steadily toward sustainability. In this sense, *sustainability* is an ethos or context from which we each strive to leave the world a little better for having been here.

The biggest challenge facing society today is the shift in paradigm demanded by ecological decision making. This transformation demands both wide-scale training and a reorientation of the direction that organizations and communities have been taking. In Chapter 5, Sinton suggests that sustainable Lean thinking offers the social transformation tools and processes needed to bring these about. The concept of using Lean thinking approaches to speed sustainable development reflects the following sustainable development principles: (1) integration of environmental criteria decision-making processes, and (2) making best use of local efforts (ORTEE 1994). The optimal utilization of local efforts and resources has emerged as one of the strongest organizing principles available to us for transforming to a sustainable society. Popular movements in the United States, like the Buy Fresh Buy Local campaign and the Business Alliance for Local Living Economies, reflect the suitability of this approach. Lean thinking applies rules that foster this approach by minimizing the movement of material and people.

The compounding actions of individuals, who, being of like mind, act in accord and can elevate the underpinnings (social values) of the current paradigm. Such actions, while often aimed at solving an immediate problem by attacking its symptoms (the case examples), form the requisite pieces that, when integrated, begin to build a systemic way of understanding and engaging sustainability (the interactive framework).

The Minnesota Regional Sustainable Development Partnerships, discussed in Chapter 6 by Draeger, Ulland, Ukaga, and Reichenbach, is a great example of community-university engagement, which not only builds on the successful land-grant tradition of applying the university research and outreach resources to local issues, but also aspires to reinvent the land-grant ideal to meet the challenges of the twenty-first century. The Sustainable Development Partnerships serves as a framework from which to build on the compact between society and public institutions. This involves local community representatives and university personnel working together to develop and implement community-based partnership projects. The ideal outcome is a partnership that provides critical resources for community-based initiatives while creating learning experiences for citizens and the university. The Regional Sustainable Development Partnerships combines citizen leadership with the research and educational resources of a public institution to foster long-term regional sustainability.

In Chapter 7, Savory and Butterfield discuss the holistic management framework, based on the assumption that sustainable development rests on a sustainable agriculture, which in turn is impossible to achieve if land is degrading. The chapter notes that the symptoms resulting from land degradation, including desertification, are exploding exponentially: increasing frequency and severity of droughts and floods even without change in rainfall; massive invasions of noxious plants and insects; poverty; violence, including abuse of women and children, as populations exceed declining resources; genocide; and, of course, migration from rural areas to cities. Because the only form of wealth that can sustain a nation is derived from the photosynthetic process, in which green plants convert solar energy to usable forms, including food, any attempt to sustain development that does not address land degradation will always be short-lived.

The point is that no matter how many millions of dollars are invested in development projects in Africa and elsewhere, there is little hope of sustained success unless two requirements are funded and met. First, education must empower the people themselves, as the only experts on their culture and situation, to actually reverse land degradation and desertification. Second, women must be educated and empowered as equals to men. As long as women remain subservient, improvements in health and rural livelihoods lead to increased family size, which the resource base cannot sustain.

Individuals who accept personal responsibility for the consequences of their thoughts, choices, and actions form the basis of sustainability. Chapter 8 on sustainable living, by Simon-Brown and Maser, is about managing ourselves. It is based on identifying the personal values, ethics, and beliefs that underlie decision making. Considering the barriers to living sustainably, examining national trends, and determining personal priorities are integral components of educating ourselves about our consumer choices. Notably, at Oregon State University and other land-grant institutions, educational efforts were focused on teaching students to professionally manage natural resources. However, as pressures of population, economics, and consumption increase, helping consumers take a thoughtful approach to understanding their cultural, economic, and environmental ethics and addressing their responsibilities as consumers of natural resources become viable educational tools.

This chapter points out that sustainable living requires us to manage the only thing we can manage—ourselves. Self-management is based on identifying the personal values, ethics, and beliefs that underlie decision making. Further, the chapter provides insight into (1) barriers to sustainable living, (2) national trends, and (3) personal priorities as integral components of educating ourselves about our consumer choices.

In the conclusion, Ukaga, Moumouni, Reichenbach, and Maser discuss participatory leadership communication as a tool for facilitating sustainable development. Rather than waiting for all the stakeholders to become interested and actively involved in a spontaneous fashion, anyone can make a positive difference by using a "growing spiral of communication" to turn an individual's sustainability idea into a community's idea and action.

Endnote

1. See http://www.brainyquote.com/quotes/quotes/e/edmundburk/00421/html (accessed on October 13, 2008).

References

ORTEE. 1994. Ontario round table on the environment and the economy (ORTEE). *Round Table Talk,* Summer.

United Nations General Assembly. 1972. *Report of the United Nations Conference on the Human Environment held at Stockholm, 5–16 June 1972*. A/CONF.48/14. New York: United Nations General Assembly.

World Commission on Environment and Development. 1987. *Our common future*. Oxford: Oxford University Press.

Acknowledgments

We thankfully acknowledge Diane Seefeldt for her help in proofreading and formatting the manuscript. We are also very grateful to our respective families for their support.

The Editors

Okechukwu Ukaga is the executive director of Northeast Minnesota Sustainable Development Partnership, University of Minnesota. He is also an adjunct professor of environmental science at the University of Minnesota–Duluth and an extension professor of sustainable development with the University of Minnesota Extension. Before coming to Minnesota, Dr. Ukaga served as the managing director of the International Institute for Sustainable Development at Colorado State University. In that capacity, he managed a variety of sustainable community development projects and worked with organizations and people from countries in Africa, South America, Asia, and the Middle East. He received a PhD in agricultural and extension education from Pennsylvania State University, an MS in education and an MBA in business from Florida A&M University, and a postgraduate diploma in agricultural economics from the University of Nigeria. He has previously taught at Colorado State University, Penn State, and Florida A&M University. His current professional and research interests are in participatory planning and evaluation of sustainable development at the local, regional, state, national, and international levels.

Dr. Ukaga has served and continues to serve on important domestic and international project teams, for example, the executive board of the Minnesota Evaluation Association, Minnesota Sea Grant College, Consortium for Sustainable Village Based Development, and Renewing the Countryside. He also served as a member of a Kettering Foundation-funded national task force on the practice of public scholarship in land-grant institutions, a member of a European Union-funded international working group on evaluation of sustainable development, and a consultant for the U.S. Agency for International Development Agribusiness Development Program (ADP) training in South Africa. His work has been applied in many parts of the world. Examples include evaluation of Regional Sustainable Development Partnerships in Minnesota, planning and evaluation of the extension service program of several universities in Indonesia, HELPO Foundation's rural development project in India, ecotourism and sustainable development in Senegal, promoting sustainable cities in New Zealand, UNICEF's country-led monitoring and evaluation systems across the world, developing a sustainability policy model for promoting cleaner production (CP) in Korea, assessment of coastal resources utilization and environmental management in Xinhui Coastal District of Gunagdong Province, China, and development of indicators for Sustainable Urban Brownfield Regeneration Integrated Management consortium case study sites in Manchester and Barking in Essex, United Kingdom.

Dr. Ukaga has written or coauthored over fifty publications, including books, a special edition of a scholarly journal, book chapters, journal articles, conference papers, and project reports. His book *Renewing the Countryside* (2001, coedited with Jan Joannides, Sara Bergan, Mark Ritchie, and Beth Waterhouse) highlights the success stories of people across Minnesota's diverse and beautiful landscape who are conserving and enhancing the state's natural and cultural resources while spurring local economic and community development. *Evaluating Sustainable Development* (2004, coauthored with Chris Maser) presents the principles and tools for participatory planning and evaluation of sustainable development. *Sustainable Development in Africa* (2005, coedited with Osita Afoaku) examines factors limiting sustainable development in Africa and offers reasoned suggestions on practical strategies for achieving development in Africa that is anchored on the values of sustainability, appropriateness, and equity. His latest book, *Sustainable Development: Principles, Frameworks, and Case Studies* (coedited with Chris Maser and Mike Reichenbach), summarizes selected sustainable development models, and includes not only theoretical pieces, but also salient case examples.

Chris Maser has spent over twenty-five years as a research scientist in natural history and ecology in forest, shrub steppe, subarctic, desert, coastal, and agricultural settings. Trained primarily as a vertebrate zoologist, he was a research mammalogist in Nubia, Egypt (1963–1964), with the Yale University Peabody Museum Prehistoric Expedition, and a research mammalogist in Nepal (1966–1967), where he participated in a study of tick-borne diseases for the U.S. Naval Medical Research Unit 3 based in Cairo, Egypt. He conducted a three-year (1970–1973) ecological survey of the Oregon Coast for the University of Puget Sound, Tacoma, Washington. He was a research ecologist with the U.S. Department of the Interior, Bureau of Land Management for thirteen years (1974–1987)—the last eight studying old-growth forests in western Oregon—and a landscape ecologist with the Environmental Protection Agency for one year (1990–1991). From 1992 to the present, he has been helping to reshape the thinking on social-environmental sustainability; to that end, he was the editor and author of a series on books on various facets of sustainability.

Today Chris Maser is an independent author as well as an international lecturer and facilitator in resolving environmental conflicts, vision statements, and sustainable community development. He is also an international consultant in forest ecology and sustainable forestry practices, and has written over 275 publications—including 28 books. His books are in academic and public libraries in every state in the United States and all but one province in Canada, as well as sixty-four other countries. (For more information, see his website: http://www.chrismaser.com/bibliog.htm.)

Michael R. Reichenbach is an educator and advocate for sustainability and environmental issues. During the environmental movement of the 1970s, he helped organize the first community recycling program in the Village of Oak Park, Illinois. He studied forestry at the University of Illinois, receiving a BS in forestry in 1978 and an MS in forestry in 1984. As a business owner working to protect natural resources and trees during development, he found ways to work with developers to achieve both economic and environmental goals. He has worked as a natural resource educator—five years for The Oregon State University and ten with the University of Minnesota Extension. He served as the associate director of the University of Minnesota Northeast Minnesota Regional Sustainable Development Partnership for one year. Currently he is a full time extension forestry educator at the University of Minnesota and is pursuing an Ed.D. in teaching and learning at the University of Minnesota Duluth College of Education.

As an educator, he organized, facilitated, or presented at more than 250 workshops, conferences, or training seminars. He provided leadership for developing and presenting educational programs on economic development, forest management, business practices, and sustainable forestry. His interest in increasing citizen involvement and understanding of sustainable development is rooted in learning about others' sense of place and how their viewpoints shape our world.

Contributors

Richard Bawden was, for many years, dean of agriculture and rural development at Hawkesbury College in Australia, during which time that institution pursued an ambitious systems-based approach to education and research. In 2000, after twenty years at that college, Dr. Bawden accepted a position at Michigan State University as a visiting distinguished university professor—from which he retired in 2007. He still teaches, consults, and is a guest lecturer on sustainable development and the need to understand our own worldviews.

Jody Butterfield is a former journalist specializing in agriculture and the environment. She is a cofounder of the Africa Centre for Holistic Management and serves on its board of trustees.

Clifford Cobb has served since 1994 as a senior fellow at Redefining Progress (RP), a nonprofit organization in Oakland, California, that advocates alternative methods of economic measurement and environmental policies based on sound economic principles. At RP, Mr. Cobb developed the genuine progress indicator (a measure of national economic welfare). Since 1997, Cobb has served in various capacities with the Robert Schalkenbach Foundation. The foundation promotes the philosophy of Henry George, best known for his advocacy of the single tax, a tax on the value of land.

Kathy Draeger is currently the statewide director for the University of Minnesota Regional Sustainable Development Partnerships (UM Regional Partnerships). This citizen-driven program connects communities to their land-grant university to advance sustainable agriculture, natural resources, renewable energy, and sustainable tourism. Ms. Draeger comes to this position having founded and run two successful environmentally based businesses: Environmental Ground, Inc., and Sustainability International, Inc., the latter of which was bought out by an engineering firm.

Charles Moumouni is an associate professor at the Department of Information and Communication of Université Laval, Quebec, and a lawyer registered with the Quebec Bar. He is the vice president of the Théophraste Network, the institutional network of francophone journalism schools. He is a member of the scientific committee of the academic agency of the Francophonie intergovernmental organization (Agence universitaire de la Francophonie). He is also the coordinator of the UNESCO–Université Laval joint training program on journalism and world heritage.

Allan Savory is president of the Savory Institute and, with his wife Jody Butterfield, a cofounder of the Africa Centre for Holistic Management of which he is chairman of the board of trustees. He pursued an early career as a research biologist and game ranger in the British Colonial Service of what was then Northern Rhodesia (today Zambia), and later as a politician, international consultant, farmer, and game rancher in Southern Rhodesia (today Zimbabwe). He went on to work, as a resource management consultant, with numerous managers, eventually on four continents, to develop sustainable solutions. Mr. Savory served as a member of parliament in the latter days of Zimbabwe's civil war and was leader of the opposition to the ruling party headed by Ian Smith. Exiled in 1979, as a result of his opposition, he immigrated to the United States. He now divides his time between Zimbabwe and New Mexico.

Viviane Simon-Brown is an extension forestry professor at Oregon State University. She coordinates the Sustainable Living Project at Oregon State University. Its mission is to improve the quality of life and reduce environmental degradation in Oregon by fostering new consumption patterns and sustainable lifestyles. Since 1998, the Sustainable Living Project at OSU has been offering off-campus programming on intelligent consumption to American adults and older youth. Over 7,500 people have participated in workshops, and over a half-million browsers have visited the website. Her international work includes research in England, Belgium, Germany, and India. She is a past president of the Association of Natural Resources Extension Professionals, and currently directs the National Network for Sustainable Living Education.

James R. Sinton is an assistant professor in the Departments of Business Administration and Sustainable Living, at Maharishi University of Management. He specializes in environmental management, Lean thinking, and sustainable living. Dr. Sinton has combined a career in management consultancy and education. He has spent over twenty years consulting for executives at large and small corporations, including IBM, Kodak, Toshiba, Procter & Gamble, and Kellogg's. This work has covered three continents: Europe, Africa, and North America.

John Talberth, an ecological economist, directs the Sustainability Indicators Program at Redefining Progress and coordinates scientific advances as well as policy applications of sustainability metrics, such as the genuine progress indicator and ecological footprint. Dr. Talberth also serves as president and consulting economist with the Center for

Sustainable Economy (CSE). At CSE, he provides technical support to nongovernmental organizations seeking to advance environmental, economic, and social sustainability through a diverse range of campaigns. Recent campaigns on which he has worked include protecting migratory birds, promoting living wage and affordable housing ordinances, ending commercial logging on national forests, and defending subsistence use of marine ecosystems in Alaska from industrial development.

Linda Ulland is the executive director of the University of Minnesota Central Region Sustainable Development Partnership, which serves an eight-county area in central Minnesota. Prior to joining the Partnership program in 2005, Ms. Ulland worked as an environmental consultant focusing on planning and project management in water resource protection and sustainable tourism issues. She has also held management positions with the Minnesota Pollution Control Agency and in the private sector with Weston Solutions in Albuquerque, New Mexico, and with Bechtel Corporation in San Francisco.

Jason Venetoulis has conducted in-depth research on the genuine progress indicator, ecological footprint analysis, and other sustainability indicators over the past decade. Dr. Venetoulis began his studies on sustainability indicators as part of Dr. Lamont Hempel's funded research at Claremont Graduate University Research Institute. During this time he traveled throughout North America interviewing sustainability program administrators at the local, county, and regional levels. His experience with the GPI began at Claremont Graduate University in the 1990s, where he conducted a local GPI for the cities of Claremont and Pomona, California, as part of his master's program. He also conducted GPIs and footprints for the San Francisco Bay Area counties as part of the efforts by the Bay Area Alliance for Sustainable Communities and national-level GPIs. He has since moved on from his work on indicators and is now working exclusively as a specialist in the solar energy field in California and British Columbia.

1

If the GDP Is Up, Why Is the GPI Down? From Normative Debate to Progressive Democratic Economics

Jason Venetoulis

CONTENTS

Part I: Normative Debate

Origins of the Counternormative to Neoliberal Economic Accounting

When offering new ideas, it is proper form to give due credit to forerunners who made such steps conceivable, a difficult task when the idea draws from a wide net. For example, the indicator of sustainable economic welfare (ISEW), the predecessor of the genuine progress indicator (GPI), was such an innovative, all-encompassing idea—drawn from myriad texts on philosophy, religion, natural sciences, social justice, feminism, economics, environmentalism, and history. A compelling, original idea may also be influenced by the personal experiences, values, and beliefs of its author(s).

Herman E. Daly and John B. Cobb Jr. spotted the first trouble in economics in the smoldering debates over relativism and positivism, community and individual, carrying capacity and perpetual growth, religion and secularism, and socialism and capitalism. They saw the rise of neoliberal economic theory as a blend of singular religous views, self-serving materialism, positivist science, and unfettered (amoral) capitalist ambitions usher in the unwaveringly pro-growth paradigm of the modern global era. And, they felt compelled to challenge the neoliberal influence in politics, religion, media, and academia by questioning, rethinking, recalculating, and offering a new way to conceptualize and measure economic progress.

John Cobb Jr. and a small team of researchers that included Cliff Cobb in Claremont, California, initially developed the ISEW. Herman E. Daly and John B. Cobb Jr. formally introduced it in the 1980s in *The Common Good* (Daly et al. 1989, 482). Despite the claims of some critics, Daly and Cobb did not oppose all economic growth or development, which was the vernacular (*lingua franca*) of twenty-first century *progress*. They did, however, masterfully identify some serious problems with the theory, methodology, and practice of the prevailing approach to economics, and offered a series of intriguing theoretical and practical alternatives.

In typical scientific fashion, a new idea has to survive a thorough vetting by peers. On the street and in the halls of power, however, acceptance of such an idea is less understandable from a scientific perspective. Should an idea be accepted, it could dramatically alter the way people view the world and, perhaps, result in something practical, such as fundamentally changing the way buildings are designed. As Bill McDonough quips, it is the innovative thinkers that we can thank for the wheel and, seemingly belatedly, wheels on luggage. While the indicator of sustainable economic welfare was well received in peer review, popular notoriety peaked with the genuine progress indicator (GPI), despite the resistance of well-endowed, organized critics' interest in maintaining the status quo.

Where traditional accounts of national income focus singularly on the economic value of monetary transactions, i.e., gross domestic product (GDP), progressive economic indicators attempt to include social, environmental, *and* economic values. The ISEW and GPI illustrate this by the inclusion of categories that are sensitive to the distribution of income distribution, as well as the value of housework and volunteerism. They also make deductions for the depletion of capital, human made and natural. As might be expected, the results are quite different from the GDP.

To illustrate, over the last forty years in the United States, the GPI has been considerably lower than the GDP. This disparity indicates that, while personal consumption may be going up, the inequities, social malaise, and degradation of nature are sending the GPI lower.

There is much to say for the wide influence the ISEW and GPI both have had, but today there appears to be a lack of success in obtaining significant real-world results. No politicians, government reports, or media outlets expound about the rise and fall of the GPI anywhere near on par with the GDP. Not surprisingly, the public is largely disengaged from the debate. And yet, under the banner of sustainability, there is an ongoing movement toward a more thoughtful and responsible approach to economic theory, accounting, and practice. Exceptional, but limited success aside, the movement's unfulfilled aspirations (at least in the case of the GPI), raise a difficult question: Why has a new measure of economic progress (or more specifically, the GPI) not been more readily accepted and applied?

What makes the question so difficult are its many variables that, when taken together, help explain the lack of broader success. In the case of the GPI, these additive explanations raise yet further questions:

- At what scale does a measure of progress matter?
- What adjustments have to be made across time and cultures?
- How should such things as the distribution of income be treated as a measure of economic progress?
- How should benefits provided by nature and the costs of polluting the atmosphere, fishing out the oceans, and clearing the Earth of trees be counted?
- In more anthropocentric terms, as a representative from the U.S. Federal Reserve put it when reviewing the San Francisco Bay Area's GPI: "Why should time spent commuting be considered a cost, when it is one of the few times of the day I enjoy myself?" (Bay Area Alliance for Sustainable Communities 2004).

The response, about doubling that time, raised doubt, but did not result in agreement about the differing experiences, perceptions, and values it raised.

Instead of using this chapter to accumulate evidence for or against divergent values, I will focus on how conflicts of normative values appeared to be dealt with in the development process of the ISEW and GPI. Though, admittedly, I may mostly agree with many of the values reflected in the GPI, my contention is that the process of development may have skewed the results in a way that has been unfavorable to broader acceptance, which is critical to moving beyond the seemingly unyielding status quo characterized by GDP and its neoliberal baggage. My intention is to contribute to the advancement of meaningful and widely used indicators of economic progress, which has much to do with politics.

Development of the ISEW and GPI

Development of the index of sustainable economic welfare (ISEW) came with an explicit acknowledgment that the "accounting system by which a nation measures its well-being is a reflection of its basic values" (Cobb et al. 1994, 9). The original impetus for openly imbuing the ISEW and gross domestic product (GDP) with value-laden theory is seemingly impossible to exact. What is clear, however, is that the originators remain open to critical discussions concerning the way in which the ISEW and GPI might be conceptualized, measured, and operationalized.

The ISEW underwent numerous reviews and revisions during its first five years of existence (Daly et al. 1989, 443). The second major proposal, critiques, and revisions were published by Clifford Cobb and John Cobb Jr. in *The Green National Product* (Cobb et al. 1994), which included eight chapters dedicated to the presentation of critiques by outside reviewers and fifty pages of subsequent changes and responses. The advances in both theory and method from this process, which approaches a critical academic (or peer) review, are apparent.[1]

For their part, the developers of the ISEW made genuine efforts to get feedback from a wide spectrum of views and backgrounds, which is not always the case in academia or the peer review process. Such an approach can be productive, given that alternative perspectives and experiences may help fill gaps in theoretical propositions and raise questions about the methodology, results, and implications, as well as checking facts. It can also help prevent ideas from getting out of hand. Despite such merits, however, this process might not be appropriate for mitigating or resolving conflicts concerning more than empirical or less than demonstrable concerns—namely, normative values.

The development of the genuine progress indicator (GPI), the successor to the ISEW, appears to be a more concentrated and strategic endeavor. Though remaining the same as the ISEW on many accounts, the GPI was debated and changed by a team of people with highly aligned political and personal agendas—though in some cases there were significantly different views. Great effort and significant resources were spent on the GPI and the launching of Redefining Progress (a nonprofit organization dedicated to the advancement of the GPI and sustainability indicators). Initial success seemed promising due to prominent coverage in well-respected national publications, course work in academia, and use by advocacy groups (Cobb et al. 1995, 59–78).

While little credit is given, the ISEW and GPI may also have spurred research on natural resources accounts at the World Bank and in the U.S. Bureau of Economic Analysis, but this work nearly stopped by the end of the 1990s. One of the more outstanding efforts to explain the GPI and stir progressive change—since Redefining Progress has moved the GPI to the back burner in recent years—has been recently carried out by Amy Taylor at the Pembina Institute in Canada. Not only were national GPI figures estimated, but also briefs were issued to policy makers, media, and the public for over twenty of the GPI categories. While the normative elements are clearly not eliminated by such a treatment, there is a "here it is, you decide for yourself" appeal to the effort. Though less rigorous than the peer review, there may be merits in dealing up front with political realities when trying to advance a new way of measuring a country's economic progress.

Although there appears to have been a sincere openness in considering other views and critical feedback in developing the ISEW and GPI, final decisions about values came from the minds of a handful of well-intentioned people. It is not clear if the underlying values conferred by the GPI are assumed to be self-evident and widely preferred, or whether they are considered compelling enough to reach the agenda of high-level, economic policy makers. What does seem evident is that challenges arising over conflicting values, perceptions, and preferences were not assailed during the process of developing and advocating the GPI.

While the GDP is still going up, the GPI's popularity seems to have stabilized and may even be waning. The lack of success (meaning broader use of the GPI) may be partially understandable because the value judgments that were made differed widely from the shared (salient) values.

Value-Laden Barriers to Success

Not surprisingly, neoliberal economists and conservative policy makers argue that the GPI is laden with assumptions based on the values of how some people think the world *should* be. One way to deal with this criticism is to offer intuitive and real-world examples about why the values underlying the GPI provide a better foundation for measuring a country's economic welfare. Before dealing with this aspect of the GPI, it is worthwhile to briefly consider the value-laden aspects of the GDP when used as a measure of economic progress.

The GDP, which measures only the dollar value of economic activity and counts all formal economic transactions as positive, was not designed as a measure of progress—as pointed out by Nobel Laureate Simon Kuznets, one of its originators. Policy makers, nevertheless, use it that way.

In promoting President Bill Clinton in 1999, his press secretary reported growth of the GDP at 4.2%—the fourth consecutive year of growth at more than 4% and the longest period of such growth since the early 1960s. In 2003, on the other side of the political aisle, President George W. Bush asserted, "We've got a consistent and effective strategy, and we're making progress.... Our third-quarter economic growth was vibrant, and that's good" (Nagourney 2003). In 2009, President Barack Obama talked about economic recovery—which at least included other aspects of the economy besides GDP growth. Still, his administration touted simple economic growth as top priority with little, if any, discussion about the qualitative health and long term sustainability of the economy.

Presidents do not deserve all the credit they take, get, or deny when it comes to the state of the economy. Political forces, societal values, environmental realities, and private sector decisions also exert some influence. In other words, the final dollar value paid for whatever is being exchanged (the gross domestic product) is influenced not only by supply and demand for land, labor, and capital, but also by social and political factors. These factors come in the form of public (political) subsidies and preferential contracts, as well as selective inaction, like that exemplified by the U.S. federal government's lack of active response to the California energy crisis in the 1990s. It may be disappointing for some economists to hear this, but considering all the factors that influence prices, I would suggest that the GDP is not really a value-free measure of free market activity—nor can it be since nothing is ever free of human valuation.

Economists implicitly deny any association between economic and noneconomic influences on the economy when they use GDP as a measure of progress in policy making and research. Least of all is there an acknowledgment of the role values play in their accounting. They may argue: "Homo economimicus has unlimited wants but no gradation of values distinct from the strength of those wants. For these reasons … whatever people desire [is] normative. The task of the economy is to meet as many of these desires as possible, whatever they be" and not to judge them (Cobb et al. 1994, 92).

Obviously, the subject of values can be a point of conflict among economists, with neoliberals assigning no lesser or higher value to different desires or actions, whether they be widely lauded, i.e., education or philanthropy, or widely disparaged, i.e., excessive (and perhaps illegal) profit taking or the proliferation of nuclear weapons. Not to assign formal criteria of evaluation to economic activity is a normative stance that is often cloaked in objective phraseology. As Neil Pert wrote and Geddy Lee sang: "If you choose not to decide, you still have made a choice" (Peart and Rush 1980).

It seems evident that policy proposals put forth by neoliberal economists and conservative politicians usually purport to maximize economic efficiency and to increase the total amount of products being consumed—often through self-serving budgetary riders with no direct reference to efficiency. The subjective value is that more efficiency/production/consumption is either paramount to the monetary bottom line or, at the very least, preferential to socioeconomic concerns not related to economic efficiency or growth. The issue of per capital income is answered with calls for market-efficient allocation of supply and demand for goods and services, as well as access to those goods and services, which is to be determined by market forces devoid of outside influences.

When it comes to biophysical systems of nature, only benefits that make someone money are added to the GDP, whereas the costs associated with the degradation and destruction of those systems are considered external to the economy they support. As Daly et al. (1989) point out, "Economists do wish to make judgments as to whether a proposed policy will improve the welfare of the people to be affected by it. But they eschew a utilitarian calculus that would add the pleasures and subtract pains. The satisfactions (economists speak of utility functions) of differing individuals are incommensurable because there is no unit of measure. Hence economists refuse to add different person's utility functions together to determine the total good to be gained."

In fact, some economists assume the GDP is the best measure of combined utility and its growth and equates more or less with economic progress. This view heavily influences the economic analyses used for policy making, teaching, and research at graduate schools. Numerous studies compare policies at the cross-national level, for example, in terms of their propensity to increase the GDP. It is not a proven fact that growth in economic activity is highly correlated with the ability to achieve the fulfillment of human desires and is thus good for society, but rather, economic growth is a value judgment somewhat akin to a worldview approximating astronomy before Copernicus. This view may also be in direct conflict with other widely shared values and new ways of looking at the world (and economy). As evidence of these assertions, one could point to the studies that have examined the relationship between quality of life, happiness, and income across time and cultures.

Studies by Richard Easterlin (2001) and the Pew Research Center (Taylor et al. 2006) suggest that happiness does not appear to be significantly correlated with economic income, especially when income levels grow in wealthier countries. At the very least, the relationship between the two appears to be stronger at lower levels of income, but is not immune from diminishing returns. This is not to say that folks do not need money to get by. The point is, the theory and corresponding algorithms put forth by neoliberal economics are steadfast in the view that increased economic income represents progress in meeting individual requirements for a quality lifestyle and thus enhances social welfare. This subjective interpretation is not necessarily supported by what happens on the ground or in people's lives.

Yet, among some economists, policy makers, media outlets, and professors there remains a reluctance to consider a different measure of progress and the GDP, which, of course, helps keep it on its pedestal. Sheer inertia in government and among conservatives in academia; biased news media; and America's overinformed, highly entertained, and often politically disengaged public may help explain why the GDP remains front and center in economic policy. This blend of conservatism and bliss presents one of the most significant barriers to the success of a new measure of economic progress—as teachers, researchers, editors/producers, and politicians protect their fiefdoms—from progress.

The Other Side of the Coin

Like the gross domestic product (GDP), the indicator of sustainable economic welfare (ISEW) and the genuine progress indicator (GPI) are rife with value judgments. There can be no doubt that the way in which an index of welfare is constructed reflects certain values, whether admitted or not. In the case of the GPI, it is recognized that the imputation of values is based on the authors' concern for the future, which may not be universally shared (Cobb and Cobb 1994, 251). The ISEW and the GPI are also much more ambitious in the sense that the GPI attempts to consider present *and* long-term social-environmental sustainability, which includes economics, whereas the GDP focuses nearly exclusively on short-term economic activity.

One manifestation of GPI's normative stance is evident in the way in which income distribution is treated. The GPI holds that equity should be the goal, not just efficiency as neoliberal economists would have it, because the GPI goes up as the distribution of income between the rich and poor equalizes. The GDP, however, does not include the distribution of income in its accounting because it's simply a measure of production or market activity. Nevertheless, it has become an extension of contemporary economics.

As such, the neoliberals contend they are being objective by focusing solely on efficiency and letting the market sort out distributional issues. For this reason, the GDP need not (indeed, should not) consider the allocation of income because, even if all the wealth goes to a small group and the masses get next to nil, the GDP is the same as it would be if it were equitably distributed. Hence, from a perspective that leans on objectivity, it is suggested that neither the GDP nor the GPI can be judged to have more merit. On the other hand, a position based on ethics (or a theory of justice) might lead to a different conclusion.

Whatever conflicts remain within the GPI as a measure of economic progress, they should probably not be left for economists to decide, whether they be neoliberal or open minded and ecologically oriented. Moreover, the important question is not whether a research team can approximate such a measure, but whether an avenue exists whereby values implicit in such an index can be expressed through a political process (Cobb et al. 1994, 8). This is a critical question because the process that determines what goes into the measure of a community's or a nation's economic progress is fundamentally a question of social values, not one of economics.

Part II: Progressive Democratic Economics

The Fourth Leg of Sustainability

Political scientist and former chair of the American Political Science Association, David Easton defined politics as the authoritative allocation of values. Sustainability is often cast in terms of environment, equity, and economy. The ISEW/GPI was the first systematic enumeration of this triad. While obviously values matter in sustainability, only recently has the fourth, distinctive leg of sustainability (politics) been given consideration in the literature.

Thomas Prugh and Robert Costanza were among the first to argue that political decision making, as a way of arbitrating values, was central to discussions about sustainability. Through political action, "people with varying interests and viewpoints can come together as political creatures and will a common environmental future. These virtues are exactly what the challenge of sustainability requires of us. Direct democracy would give us both better eyes to see the environmental problems we face and superior political means to address them" (Prugh et al. 2000).

Environmental concerns aside for the moment, the current political system in America and other major democracies is less informed, thoughtful, and representative than it could be if a direct, deliberative democracy was the norm. Although this point is not beyond contention, as will be discussed in a moment, consider a rather elaborate definition of (deliberative) democracy.

Donatella della Porta (2005, 73–94) identified seven characteristics of a deliberative democracy:

1. *Preference (trans)formation.* "A process through which initial preferences are transformed in order to take into account the points of view of the others" (Miller 1993, 75). In fact, "deliberative democracy requires the transformation of preferences in interaction" (Dryzek 2000, 79).
2. *Orientation to the public good.* In this model of democracy, "the political debate is organized around alternative conceptions of the public good"; above all, it "draws identities and citizens' interests in ways that contribute to public building of public good" (Cohen 1989, 18–19). Democratic self-restraints should prevent people from pursuing self-interest (Miller 1993, 195). A deliberative setting facilitates the search for a common end or good (Elster 1998).
3. *Rational argument.* The force of a better argument convinces people in a deliberative democracy. In particular, deliberation is based on multiple contributors to a discussion, wide opportunities for interaction, confrontation on the basis of rational argumentation,

and reciprocal listening (Habermas 1981:1996). In this sense, deliberative democracy is discursive.

4. *Consensus.* All participants approve decisions—in contrast to a majority rule democracy, wherein decisions are legitimized by votes. In this sense, deliberative democracy is consensual.
5. *Equality.* It "requires some forms of apparent equality among citizens"; in fact, deliberation takes place among free and equal citizens (as "free deliberation among equals") (Cohen 1989, 74–92). At least, "all citizens must be empowered to develop those capacities that give them effective access to the public sphere," and "once in public, they must be given sufficient respect and recognition so as to be able to influence decisions that affect them in a favorable direction." Deliberation must not only exclude coercion but also an unequal weighting of influence by representatives of organizations of different size or social status (Bohman 1997, 321–48).
6. *Inclusiveness.* All citizens who will be affected by the decisions to common problems must be included in the process and able to safely express their diverse perspectives. Deliberation (or even communication) is based upon the belief that, while not giving up my perspective, I might learn if I listen to the other. In this sense, deliberative democracy is linked to the concept of associational democracy (Young 1996, 120–35).
7. *Transparency.* In Joshua Cohen's definition, a deliberative democracy is "an association whose affairs are governed by the public deliberation of its members" (Cohen 1989, 74–92). Public deliberation can "replace the language of interest with the language of reason" (Elster 1998, 111).

So, deliberative democracy is, at least in part, contingent on conditions of equality, inclusiveness, and transparency, a communicative process based on the strength of the argument that transforms individual preferences into consensual decision making oriented to the public good (Della Porta 2005, 73–94).

A potential benefit of deliberative democracy is that latent issues of significance can be introduced into a discussion of sustainability and be given thoughtful consideration by a wider group than those who are already familiar with them. This process facilitates learning as values are discussed, debated, and perhaps brought to consensus. The broader public benefit, however, could be a more representative distribution of the decision-making power in politics.

Jeffersonian democratic aspirations posit a preference for an inclusive political system, as opposed to one based on efficiency (i.e., strongly centralized government). The liberty promised by Jeffersonians through local control remains an important, conservative value in America that also has some support among progressives on the left. In its ideal, Jeffersonian democracy might be thought of as the political equivalent of free market economic theory. Both prefer a "place" where choices are freely made by individuals.

That notwithstanding, Madison's views captured the constitutional convention, and today the delegation of authority up the proverbial chain of command is widely accepted in America and other democracies, though perhaps not entirely supported. Of course, Madison's notion of democracy is less representative than the alternative Thomas Jefferson worked for. Yet today, even among those who advocate highly democratic political systems as the only viable check to a (predominantly) capitalist economy, there is an understanding that, like free market ideals versus actual practice, democracy has its limits. It is *not* a panacea.

Dating back to Aristotle, who may have been predated by the Iroquois, debates over the best form of democracy have been around for thousands of years. Today, protests against highly participatory democratic efforts are brought to bear by neoliberal economists who favor political systems that value efficiency over representation and social effectiveness. In the real world, the case of Chile in the 1970s poses significant challenges to this perspective, as do current events in Iraq. Although some may slander democracy based on standard values (e.g., advocating efficiency over more representation), there seems little point for such occasions except to maintain the status quo.

There are also practical challenges to meeting the goals of deliberative democracy. The first issue concerns the transformation of values into economic policy (and prices). To paraphrase Oscar Wilde, some folks know the price of everything but the value of nothing. A policy-making process that offers informed voters an opportunity to address economic questions may help sort out the general details of a new measure of economic progress. For example, a voter might be asked if a production cost that results in pollution or higher costs in health care associated with pollution should be counted as positive, negative, or not at all in an economic indicator. On this point, I am not inclined to delve deeply into the operational mechanics of democracy as a way to develop a new economic measure of progress, though I will offer a few thoughts.

One way to approach progressive economic accounting, such as the GPI, is to begin with an introduction of potential categories: income, distribution, social and environmental costs. Discuss the alternative perspective and normative assumptions related to each, and then move to discussions about ways to estimate them. This information could be collated and presented by a politically diverse group to interested organizations, policy makers, and the broader public via media outlets. From here, polls and surveys could be used to gather feedback and develop a second round of questions in which participants are asked to evaluate and rank the prices (economic values) of the respective categories. While such a process is considerably more time-consuming than a team of economists and political strategists developing their own approach, it is likely to have several worthwhile advantages.

First, given an appropriate scale, the outcome of a deliberative, democratic process is likely to represent widely shared economic concerns. This may help neutralize anxieties associated with the subjectivity of normative stances. Second, there is likely to be greater support from the public and policy makers. Third, if participatory evaluations and rankings are done periodically, it may better reflect the dynamic aspects of what is important in terms of economic progress. And finally, it might help displace the GDP as a measure of progress, though not negating its use as a measure of economic production with agreed-upon qualifications.

The other practical difficulty with direct, deliberative democracy is referred to as the scaling up (Friedman 2006). In other words, deliberative democracy can be more manageable and representative at the local level than at the national level, where an indicator of progress has more significant economic and policy implications. In dealing with the challenges of scaling up, a number of promising efforts have either been undertaken or are under way. Public Agenda, Stanford Professor James Fishkin's efforts in "deliberative polls" (Fishkin 1991, 133), and Viewpoint Learning's Choice Dialogues (www.viewpointlearning.com) are leading examples (Viewpoint Learning 2008).

It is acknowledged even among practioners in the field, however, that these efforts have fallen short of meeting the ideals of widespread participation. While the ideal may remain out of reach, changes to the political system that facilitate democratic participation and equalize political relations are still worthwhile. As Thomas Purgh has suggested, it is not a matter of perfecting democracy, but rather of using democracy to make sure we fare well in terms of sustainability. Further, and more to the point, as Cobb and Cobb Jr. assert, the intent is to develop a measure of economic progress that is meaningful to those it affects, in contrast to one imposed by an analyst's desire to compare one person, community, or country to another (Cobb et al. 1994).

Short of the democratic revolution, which instituting a direct democracy would require, the popular initiative process that exists in many political jurisdictions could provide a vehicle for developing measures of economic progress. Initiatives hold some promise for dealing with scale issues as well, albeit in an incremental way. While potentially not as representative as political choice made through a highly democratic process, initiatives do offer a way in which people's concerns and proposed remedies can be vetted in the popular arena of the polis.

Although slowing down the entire process, community-level policy initiatives that institutionalize sustainable economic indicators could precede those of the state and region. These, in turn, could inform the state-level process, which could then inform the regional-level process. Of course, there is no national initiative process in the United States or other major democracies, but there are ways to get issues on the national agenda that do not rely on campaign contributions or the attraction of fleeting media attention.

Perhaps the biggest problem with the initiative process is that most democracies lack one. This aside, initiatives tend to require a significant amount of collective action and resources, both of which are scarce political currency. Moreover, the outcome of an initiative process is likely to be less representative than one coming from a democratic revolution. And finally, initiative-driven laws, when passed by voters, are often challenged in court (the least democratic branch of government) by those whose interests and values lose to the voting populous. True, such an approach does not deal directly with the problems of deliberative democracy, but a proposed policy could help spur civic engagement and stimulate existing democratic processes in the development and adoption of a new, widely accepted and utilized indicator of economic progress.

Power to the People

For those less inclined to direct political action and developing new economic indicators, but still interested in sustainability, allow me to digress briefly in a discussion about the political, economic, and environmental implications of energy, which it seems to me relates well to our topic.

There are practical, market-based alternatives for making progress on the economic, environmental, political, and social justice fronts of sustainability. Along these lines, perhaps one of the most substantial commitments that a business, nonprofit organization, or homeowner can make is to purchase a home- or business-based renewable energy system. Solar panels on the roof of a home or a wind turbine at the factory can be initially expensive, but the payoffs are a stronger local economy, less pollution, and devolution of political power. By producing energy locally, less money is exported to pay for imported energy. In addition, renewable energy generates less pollution than does fossil fuel or nuclear sources. And, since no one owns the sun, those who own or control today's major sources of energy in the private and public sector may lose some political and economic power.

At US$20,000 to 50,000, however, the initial price of a two-kilowatt solar energy system and energy-efficient lighting and appliances might be too high a monetary price for many who would prefer the economic, environmental, and political benefits of renewable energy. This cost differential exists, in part, because the latter is offered at its marginal price (i.e., ten cents per kilowatt), while renewable energy systems are typically available only in large packages with the entire cost due immediately, which does not negate the fact that energy from solar panels will cost nearly two-thirds less over twenty-five years—and that assumes energy prices do not go up. For college campuses, manufacturing facilities, and homes that are expected to be around for more than ten years, the math on renewable energy seems to make sense because a full return on dollars spent now takes less than a decade. Thereafter, resources that went to energy can go to such things as lower tuition, better wages, and home improvements.

Recently, several small, private companies have emerged to begin filling the desire for even lower-cost renewable energy. Notable among these efforts is Native Energy, which sells renewable energy to the public on a per watt basis and uses the revenue to install wind energy systems on Native American reservations (nativeenergy.com). Accelerate Solar (a nonprofit I am trying to get off the ground in my very little spare time) does similar work, but with a focus on solar energy for low- and moderate-income homeowners (acceleratesolar.com). Internationally, the nonprofit Sun Electric Light Fund (self.org) has advanced the use of renewable energy by securing support from foundations and donors and using funds to bring water and light to health care and community centers in rural villages. Contributions to such organizations have many of the same advantages as purchasing an entire system for your home or business, although the benefits are less direct.

For those not interested in a (velvet) democratic revolution and too busy to deal with initiatives, a viable alternative is to vote for politicians who will work for sustainability.

At the Closing Bell

While the GDP is reported quarterly, stock market figures are reported daily. When the housing market moves 1/16th of a percent, the news hounds bellow. Interest rates, consumer confidence, and currency exchange rates are also gaining notoriety—contributing yet further to the entrenchment of the "growth is always good camp." While the GDP is where the culmination of economic output is added up, the prospects for achieving democratically informed, progressive economics dims with the rise of each indicator of growth.

In the final tally, using democracy to manage the economy and develop measures of economic progress does not address some difficult problems that arise from a strict monetary focus. The essence of the problem is in considering the value of everything in terms of money—nature and social relations in particular. What does it mean, for example, when the price (cost) of emitting a ton of carbon goes up? While it may reflect the increasing ecological costs of carbon emissions as they relate to climate change, the price cannot tell you that the global climate has become highly destabilized, which seems, at the very least, just as important as the price of carbon credits or oil, for that matter. The economy, as so many have already stated, is dependent on nature, which includes people. The need for indicators and, more importantly, actions that ensure ecological integrity can only be ignored by accepting the potential devastating peril now facing all life. Nevertheless, there remains great potential for genuine progress to the extent that true, participatory democratic processes can provide a way to move beyond entrenched debates about the economic toll of environmental protection and social programs.

Acknowledgment

Much great thanks to Clifford Cobb for reviewing and commenting on an early draft of this chapter, and his outstanding work in progressive economics over the years.

Endnote

1. The peer review process includes the submission of new ideas or new treatments of old data, for example, followed by critiques from independent knowledgeable reviewers, and then, if deemed worthy, an opportunity for authors to respond by making changes or counterarguments before reconsideration for publication.

References

Bay Area Alliance for Sustainable Communities. 2004. A review of the Bay Area's genuine progress indicator. Personal communication.

Bohman, James. 1997. Deliberative democracy and effective social freedom: Capabilities, resources and opportunities. In *Deliberative democracy: Essays on reason and politics*, ed. James Bohman and William Rehg, 321–48. Cambridge, MA: MIT Press.

Cobb, Clifford W., and John B. Cobb. 1994. *The green national product: A proposed index of sustainable economic welfare*, 251. Lanham, MD: University Press of America.

Cobb, Clifford W., John B. Cobb, and Human Economy Center. 1994. *The green national product: A proposed index of sustainable economic welfare*. Lanham, MD: University Press of America.

Cobb, Clifford W., Ted Halstead, and Jonathan Rowe. 1995. If the GDP is up, why is America down? *Atlantic Monthly*, October.

Cohen, Joshua. 1989. Deliberation and democratic legitimacy. In *The good polity: Normative analysis of the state*, ed. Alan P. Hamlin and Philip Pettit, 74–92. Oxford: Blackwell.

Daly, Herman E., John B. Cobb, and Clifford W. Cobb. 1989. *For the common good: Redirecting the economy toward community, the environment, and a sustainable future*. Boston: Beacon Press.

Della Porta, Donatella. 2005. Making the polis: Social forums and democracy in the global justice movement. *Mobilization* 10:73–94.

Dryzek, John S. 2000. *Deliberative democracy and beyond: Liberals, critics, contestations*. New York: Oxford University Press.

Easterlin, David. 2001. Income and happiness: Toward a unified theory. *Economic Journal* 111:465–83.

Elster, Jon. 1998. Deliberation and constitution making. In *Deliberative democracy*, ed. Jon Elster, 97–122. New York: Cambridge University Press.

Fishkin, James S. 1991. *Democracy and deliberation: New directions for democratic reform*. New Haven: Yale University Press.

Friedman, Will. 2006. Deliberative democracy and the problem of scope. *Journal of Public Deliberation* 2 (no. 1, article 1).

Habermas, Jürgen. 1981. *Theorie des Kommunikativen Handelns*. Habermas, Juergen. 1996. *Between facts and norms: Contribution to a discursive theory of law and democracy*. Cambridge: MIT Press. 1. Aufl ed. Frankfurt am Main: Suhrkamp.

Miller, David. 1993. Deliberative democracy and social choice. In *Prospects for democracy: North, south, east, west*, ed. David Held, 74–92. Stanford, CA: Stanford University Press.

Nagourney, Adam. 2003. Campaign tactics being reversed as events shift. George W. Bush quote in the *New York Times*, November 4.

Neal Peart and Rush. 1980. *Freewill*, ed. Geddy Lee, Alex Lifeson, and Rush. Vol. record album, Permanent Waves Mercury-Polygram.

Prugh, Thomas, Robert Costanza, and Herman E. Daly. 2000. *The local politics of global sustainability*. Washington, DC: Island Press.

Taylor, Paul, Cary Funk, and Peyton Craighill. 2006. *Are we happy yet? A social trends report*. Washington, DC: Pew Research Center.

Viewpoint Learning. 2008. Available from http://www.viewpointlearning.com/ (accessed April 13, 2008).

Young, Iris Marion. 1996. Communication and the other: Beyond deliberative democracy. In *Democracy and difference: Contesting boundaries of the political*, ed. Seyla Benhabib, 120–35. Princeton, NJ: Princeton University Press.

2

Sustainable Development and the Genuine Progress Indicator: An Updated Methodology and Application in Policy Settings

John Talberth and Clifford Cobb

CONTENTS

The Challenge of Sustainable Welfare Measurement

On October 28, 2005, the following headlines appeared in leading newspapers throughout the United States:

> "GDP Muscles Through"
>
> "Economy Brushes Off Storms and Expands by 3.8 Percent in 3Q, Beating Estimates"
>
> "The U.S. Economy Shook Off Headwinds from Hurricanes Katrina and Rita to Grow at a Faster-than-Expected 3.8 Percent Annual Rate in the Third Quarter, a Commerce Department Report Showed Friday" (Reuters)

Perhaps no headline in recent history does a better job of illustrating why our nation's most trusted measure of economic performance is so woefully out of sync with people's everyday experiences. In one swoop, these headlines dismissed the inequitable and catastrophic social, environmental, and economic toll associated with 1,836 preventable deaths; over 850,000 housing units damaged, destroyed, or left uninhabitable; disruption of 600,000 jobs; permanent inundation of 118 square miles of marshland; destruction of 1.3 million acres of forest; and contamination caused by millions of gallons of floodwaters tainted by sewage, oil, heavy metals, pesticides, and other toxins as irrelevant to the U.S. economy.[1]

Few would dispute the fact that gross domestic product (GDP) fails as a true measure of economic welfare. For decades, many economists have acknowledged that the GDP has fundamental shortcomings. "GDP is not a measure of welfare," wrote William Nordhaus and James Tobin, prominent economists at Yale University in the early 1970s (Nordhaus and Tobin 1972). The GDP is simply a gross tally of everything produced in the United States—products and services, both good and bad. In fact, in a 1934 report to Congress, GDP's chief architect, Simon Kuznets, cautioned that "the welfare of a nation can scarcely be inferred from a measurement of national income" (Kuznets 1934, 7).

Despite these cautions, GDP retains its prominent catchall role for our collective human well-being due, perhaps, to little consensus on a suitable replacement or, more fundamentally, if well-being can be quantitatively measured at all. Nevertheless, efforts to find replacements are critical since GDP is the basis for important public policy decisions; i.e., decisions whose outcomes are predicted to increase the GDP fare better than those shown to restrict the GDP. Recently, growth of the GDP was a prominent justification for highly controversial tax cuts on capital gains, whereas efforts to secure long-overdue increases in the federal living wage have been thwarted by persistent forecasts of gloom and doom with respect to jobs and economic growth (Foertsch 2006; Roth 2005).

In this chapter, we present an update to the genuine progress indicator (GPI)—one of the first alternatives to the GDP vetted by the scientific community and regularly used worldwide by government and nongovernmental organizations alike. The GPI and its variants, such as the index of sustainable economic welfare, were conceived as a way to measure changes in national economic welfare with a single, aggregate index. The GPI considers households as the basic building block of a nation's welfare, and thus begins its accounting exercise with expenditures on personal consumption. To this, the GPI adds benefits associated with activities that enhance personal welfare, such as parenting, housework, volunteering, and higher education, as well as the services that flow from household capital and public infrastructure. The GPI then deducts costs associated with pollution, loss of leisure time, auto accidents, and destruction of natural capital, the accumulation of international debt, and the depletion of resources. The end result is an index that attempts to measure our collective welfare in terms of principles of sustainability drawn from the economic, social, and environmental domains.

By differentiating between economic activity that diminishes social-environmental capital and that which enhances such capital, the GPI is designed to measure sustainable economic welfare rather than solely economic activity. In particular, if the GPI is stable or increasing in a given year, the implication is that stocks of social–environmental capital, on which all goods and services depend, will be at least as great for the next generation as for the present one. But on the other hand, if the GPI is falling, it implies that the economic system is eroding those stocks and limiting the next generation's prospects. The remainder of this chapter is organized as follows.

Section 1 gives a brief overview of the evolution of the GPI framework. Venetoulis (in Chapter 1) also discussed in more detail the evolution of the GPI, its normative aspects, and its alignment with popular notions of social, economic, and environmental sustainability. As such, we do not replicate that discussion herein. Instead, we concentrate on theory, methods, and policy implications. Section 2 is a discussion of the disconnection between the GDP and true economic welfare, and how the GPI responds to these defects. Section 3 is a review of the GPI's theoretical underpinnings and its critiques. Section 4 is an updated explanation of GPI methodology. Section 5 presents results of the 2006 update and main findings. Section 6 demonstrates how the GPI can be used to inform public policy debates using globalization, tax cuts, and sprawl as examples. Finally, concluding thoughts and directions for future research are set forth in Section 7.

Section 1: Evolution of the Genuine Progress Indicator Framework

During World War II, gross domestic product (then gross national product) accounts were introduced to measure the capacity of wartime production (Cobb et al. 1995). Since then, the GDP has become the world's most ubiquitous indicator of economic progress. It is widely used by policy makers, economists, international agencies, and the media as the primary scorecard of a nation's economic health and well-being. Yet, as we know from its creator, Simon Kuznets, the GDP was never intended for this role (Kuznets 1934). It is merely a gross tally of products and services bought and sold, with no distinctions between transactions that enhance our well-being and those that diminish it. Instead of distinguishing costs from benefits, productive activities from destructive ones, or sustainable ones from unsustainable ones, the GDP—by definition—simply assumes that every monetary transaction adds to social well-being. In this way, needless expenditures triggered by crime, accidents, contamination from toxic wastes, preventable natural disasters, and corporate fraud count the same as socially productive investments in housing, education, health care, sanitation, or mass transportation. It's as if a business tried to assess its financial condition by simply adding up its "business activity" by lumping income and expenses, assets and liabilities together in a single column.

Moreover, GDP ignores everything that happens outside the realm of monetized exchange, regardless of its importance to our well-being. The crucial economic functions performed in a household and countless volunteer hours go entirely unnoticed, as do ecosystem services, such as flood control, water storage and filtration, carbon sequestration, soil formation, and maintenance of genetic diversity. As such, the GDP devalues welfare-enhancing activities, such as child and elder care, mentoring, or ecological restoration. In fact, the GDP ignores the entire informal, or noncash, economy—a significant component of the overall exchange system worldwide. In a 2002 analysis, the International Monetary Fund reported that the global value added by the informal economy had reached a "remarkably large amount"—up to 44% of the GDP in industrialized nations, 30% in transition economies, and 16% in Organization for Economic Cooperation and Development (OECD) economies (Schneider and Enste 2002). In the United States, the size of the informal economy is not systematically surveyed, but conservative estimates place its current size at 9% of the official GDP, involving up to 25 million Americans (Barber 2003).

Because the GDP fails to properly distinguish between welfare-enhancing and welfare-degrading expenditures and ignores nonmonetized costs and benefits, including all informal exchanges, using the GDP as a barometer of overall well-being leads to some perverse results. Consider these:

- The GDP increases with polluting activities and then again with cleanups. Pollution is a double benefit to the economy since the GDP grows when we manufacture toxic chemicals, when they are purchased and used, and again when we are forced to clean them up.
- The GDP is boosted by crime. Each year, Americans incur nearly $40 billion in crime-related costs in the form of lost and damaged property, as well as the replacement or repairs to said property. And then there are the crime prevention expenditures on locks, alarms, and security systems. The GDP counts these needless expenditures as an economic gain, implying that crime is good for our well-being.
- The GDP is oblivious to gross inequality. If a billionaire spends $10,000 on aphrodisiacs made from the horn of an endangered rhinoceros, it counts the same as $10,000 spent by a New Orleans flood victim on bare essentials as far as the GDP is concerned. As long as overall expenditures are increasing, the GDP will grow, even if the increase is entirely attributable to conspicuous habits of consumption by the wealthy even as others wallow in abject poverty.

Section 2: How the GPI Attempts to Correct These Deficiencies

Beginning with the seminal work of Daly and Cobb (1989), there have been several attempts to develop alternative accounting systems for national productivity or well-being that address aforementioned deficiencies. Collectively, these systems measure what is commonly referred to as "green" GDP. Major objectives of these green GDP accounting systems are to provide a more accurate measure of welfare and to gauge whether or not an economy is on a sustainable path (Hanley 2000). Two of the most popular green GDP systems are the index of sustainable economic welfare and the genuine progress indicator. While methodologies differ somewhat, green GDP accounting systems all involve three basic steps (Stockhammer et al. 1997; Neumayer 2000).

Computation usually begins with estimates of personal consumption expenditures, which are weighted by an index of the inequality in the distribution of income to reflect the social costs of inequality and diminishing returns to income received by the wealthy. Additions are made to account for the nonmarket benefits associated with volunteer time, housework, parenting, and other socially productive uses of time, as well as services related to both household capital and public infrastructure. Deductions are then made to account for purely remedial expenditures, such as those related to pollution or automobile accidents, as well as costs that reflect the undesirable side effects brought on in the name of economic progress. Deductions are also made at this stage for costs associated with degradation or depletion of natural capital incurred by existing and future generations.

In this way, green GDP systems correct the deficiencies of the GDP by incorporating aspects of the nonmarket economy, separating welfare-enhancing benefits from welfare-detracting costs, correcting for the unequal distribution of income, and distinguishing between sustainable and unsustainable forms of consumption. Applications of these new accounting systems in Australia, Austria, Chile, Germany, Italy, the Netherlands, Scotland, Sweden, the United Kingdom, and the United States provide compelling evidence of a widening gap between traditional and green GDP. What this means is that economic growth in these nations is increasingly being offset by the social and environmental costs of that growth so that true welfare is stagnant or on the decline (Max-Neef 1995).

Section 3: Theory and Critiques

To understand the theoretical foundations for the GPI, it is important to clarify exactly what the GPI is actually measuring. Summarizing the literature, Asheim (2000) identifies three kinds of measurements green GDP accounts like the GPI attempt to undertake: (1) welfare-equivalent income, (2) sustainable income, and (3) net social profit. Welfare-equivalent income refers to the degree one's welfare is either enhanced by or harmed by the items one purchases (Fisher 1906). Paraphrasing Fisher, Lawn (2003, 111) explains, "The national dividend consists not of the goods produced in a particular year, but of the services enjoyed by the ultimate consumers of all human-made goods." In recognition of the fact that economic process involves many "irksome" activities, which precludes the improvement of welfare with increasing levels of consumption, the concept of welfare-equivalent income should be thought of in a net sense, which means that harmful aspects of consumption are deducted from its beneficial aspects (Lawn 2003). To accomplish this, green accounts first isolate expenditures related to personal consumption by removing money spent purchasing, maintaining, or replacing durable goods (such as cars, washing machines, and televisions) and then make a series of additions or deductions to reflect benefits and costs associated with that consumption overlooked by GDP.

Sustainable income refers to the basic notion of income as defined by Sir John Hicks in *Value and Capital* (1947, 179), where he maintains that "we ought to define a man's income as the maximum value which he can consume during a week, and still expect to be as well off at the end of the week as he was at the beginning." Although Hicks' concept of income renders it sustainable by definition, few noneconomists know about Hicks or his definition of income. Therefore, sustainable income is more readily understood by the general public and is thus more relevant. To arrive at an adequate measure of sustainable income, green accounts deduct from the GDP depreciation of stockpiles of both human-created and natural capital and certain expenditures (i.e., on security systems of all types) made to defend us from some of the undesirable side effects of economic growth (Daly and Cobb 1989).

Net social profit is a measure of policy effectiveness in that it can be positive, which indicates the proposed policy is welfare enhancing, or negative, thereby indicating that its social costs exceed its benefits. Net social profit analysis is simply an expanded form of cost-benefit analysis that uses welfare-equivalent or sustainable income rather than GDP. Thus, using green accounts in net social profit analysis provides a measure of the welfare or sustainability as it might be affected by policy changes (Asheim 2000). In particular, net social profit is the difference between green GDP with and without a particular policy change.

Although the genuine progress indicator can be of use in calculating welfare-equivalent income, sustainable income, or net social profit, in aggregate, it falls squarely under the concept of welfare-equivalent income—since it attempts to measure the net welfare-enhancing income households derive from their various types of consumption. However, it only counts the portion of net welfare-enhancing income that is sustainable, or derived from stable or increasing stocks of human-created and natural capital. While certainly a more accurate measure of true welfare than the GDP or green GDP accounts rooted in notions of sustainable income, the methodological objectivity of net welfare-enhancing measures (such as the GPI) is necessarily much less clear because they necessitate value judgments over what does and does not constitute welfare-enhancing forms of consumption, what costs and benefits are added or deducted from such consumption, and how these costs and benefits ought to be measured. To this end, Venetoulis (in Chapter 1) provides a thoughtful exposition on how current normative values that permeate the GPI may, in fact, be hindering its more widespread use.

Despite its roots in standard economic theory and widely shared principles of sustainable development, the GPI is not without its detractors. Criticisms have been leveled at its theoretical foundations, components, and methods of calculation. Many of the concerns were addressed during the formative years of the GPI. In their *The Green National Product: A Proposed Index of Sustainable Economic Welfare*, Cobb and Cobb published a series of critical essays that described how those criticisms were dealt with in the revised GPI accounts (Cobb and Cobb 1994). It is not our intent to revisit those debates. Instead, we focus here on lingering criticisms.

Neumayer (1999), Dietz and Neumayer (2006), and Lawn (2003, 2005) have engaged in the most visible dialogue in the recent literature. Neumayer and others argue that it is theoretically "not possible to combine an indicator of current welfare with an indicator of sustainability" because costs associated with depletion of nonrenewable resources and other forms of natural capital incurred by future generations make little difference to current welfare (Dietz and Neumayer 2006, 189). Deductions for the depletion of natural capital are inconsistent with the net welfare-enhancing notion of income the GPI purports to measure. In response, Lawn (2005) maintains that it is entirely appropriate to assign a cost to natural capital depletion because the GPI is interested in *sustainable* economic welfare, not economic welfare per se. As such, it is important to deduct from national income accounts that portion of economic activity that cannot be sustained because it is based on depletion of natural capital.

Another theoretical flaw is the fact that, while the GPI purports to be based on the principle of strong sustainability—which considers natural capital irreplaceable—it in fact measures weak sustainability by treating natural and human-built capital as substitutes. In particular, since the GPI measures the loss of both natural and human-built capital separately, if natural capital is depleted, the costs of doing so can be masked by substitution of human-built capital of equal or greater value. According to Neumayer (1999, 93), "ironically, the Index of Sustainable Economic Welfare does not measure strong sustainability, but weak sustainability at best since it assumes perfect substitutability among different forms of capital."

The most important critique is that the GPI is arbitrary in what components it includes or implicitly excludes as contributors to or detractors from welfare (Neumayer 1999). For instance, the GPI corrects for income inequality but does not include corrections for the degree of political freedom or degree of equality between the sexes. The inclusion of almost every disservice item (i.e., commuting costs, loss of leisure, noise pollution) has been challenged because it is unclear whether or not these costs have already been factored into household and worker decisions (Lawn 2005; Rymes 1992).

In terms of calculation methods, Dietz and Neumayer (2006) take issue with four components: (1) the valuation of the depletion of nonrenewable resources, (2) the cumulative cost of long-term environmental damage, (3) the adjustment of expenditures for personal consumption based on income inequality, and (4) the deduction of remedial expenditures. These critiques involve the precision of the methods used in calculation, not the basic components. There have also been a number of criticisms made concerning the sources of data used for calculating individual GPI subaccounts. As described by Lawn, the lack of appropriate data for many GPI components forces GPI practitioners to "make heroic assumptions" to ensure that the values of these items are decent approximations of their correct value (Lawn 2005, 199).

Despite these lingering theoretical and methodological issues, the most outspoken recent critic of the genuine progress indicator (GPI) and index of sustainable economic welfare (ISEW) has concluded:

> … the ISEW's focus on comprehensive current welfare is laudable. Indeed, the emerging sustainable consumption discourse gives the ISEW renewed salience because, according to some, the task of making society consume more sustainably is in large part a question of separating out those things that we consume that make us "happier" and those that don't or even make us less happy. (Dietz and Neumayer 2006, 190)

Section 4: An Updated GPI Methodology

The GPI is derived from twenty-six separate time-series data columns spanning the years 1950–2004. Due to delays in government reporting, there is a two-year time lag in publishing GPI accounts. Table 2.1a and b describes deductions and contributions to the GPI and reports its 2004 value. In this section, we review calculation, for those columns where we have changed the methodology or made significant updates to the underlying data. For each of these items, we briefly describe the rationale for inclusion, sources of data on which we rely, and general methodology for our calculations. The methodology presented here represents a significant update to that used since the late 1990s, as described by Anielski and

TABLE 2.1A
GPI Contributions (2004)

Item	GPI Contributions		Amount (Billions)
	Personal consumption expenditures		$7,588.60
D	Weighted personal consumption expenditures (adjusted for inequality)	+	6,318.41
E	Value of housework and parenting	+	2,542.16
F	Value of higher education	+	827.98
G	Value of volunteer work	+	131.30
H	Services of consumer durables	+	743.72
I	Services of streets and highways	+	111.55
Z	Net capital investment (positive in 2004, so included in contributions)	+	388.80
	Total positive contributions to the GPI		**$11,063.92**

TABLE 2.1B
GPI Deductions (2004)

Item	GPI Deductions		Amount (Billions)
J	Cost of crime	–	$34.22
K	Loss of leisure time	–	401.92
L	Costs of unemployment and underemployment	–	176.96
M	Cost of consumer durable purchases	–	1,089.91
N	Cost of commuting	–	522.61
O	Cost of household pollution abatement	–	21.26
P	Cost of auto accidents	–	175.18
Q	Cost of water pollution	–	119.72
R	Cost of air pollution	–	40.05
S	Cost of noise pollution	–	18.21
T	Loss of wetlands	–	53.26
U	Loss of farmland	–	263.86
V	Loss of primary forest cover	–	50.64
W	Depletion of nonrenewable resources	–	1,761.27
X	Carbon dioxide emissions damage	–	1,182.82
Y	Cost of ozone depletion	–	478.92
AA	Net foreign borrowing (positive in 2004, so included in deductions)	–	254.02
	Total negative deductions to the GPI		**$6,644.83**
AB	**Genuine progress indicator 2004 (2000 dollar values)**		**$4,419.09**
AC	**Genuine progress indicator per capita 2004 (2000 dollar values)**		**$15,035.65**

Rowe in *The Genuine Progress Indicator* (1999). Methodologies for the items not described can be taken from that volume. Unless otherwise noted, all figures are reported in 2000 dollar values. We encourage readers to contact the authors for a more detailed explanation and for the most up-to-date reference information for the time-series data sets.

GPI Contributions

Item E: Value of Household Work and Parenting

Work performed in households is more essential than much of the work done in offices, factories, and stores. Yet most of this goes unaccounted for in the calculation of national income. While housework and parenting of the stay-at-home moms or dads count for nothing in the GDP, commercial child care in the monetized "service sector" adds to the GDP. Other unpaid household labor, such as physical maintenance of the housing stock (from cleaning to light repairs), also constitutes valuable economic activity.

The calculation of the value of household labor in the GPI is derived from the work of economist Robert Eisner, past president of the American Economics Association. Eisner first derived estimates of annual hours spent performing relevant household tasks from time use studies conducted by the Michigan Survey Research Center in 1965, 1975, and 1981. He then treated the value of an hour of housework as equivalent to the amount that a family would have to pay to hire someone to do equivalent work in their home. This yields an estimate of the total annual value of household work (Eisner 1985). Our GPI update incorporates three new data points: one from the final Michigan Survey Research Center study in 1985 and two from the Bureau of Labor Statistics (BLS) American Time Use Surveys (ATUS) of 2003 and 2004. We interpolated for the years in between by using a regression on the years 1981, 1985, 2003, and 2004. Each data point was incorporated slightly differently.

For the 1985 estimate, we replicated Eisner's methodology as closely as possible. Starting with raw data from the Michigan survey, we calculated the number of hours of household work performed by each of four groups: employed men, unemployed men, employed women, and unemployed women. We then multiplied those numbers by each group's respective U.S. population to calculate the total number of hours of household work performed: 235 billion. In 1985, the work was valued at $7.14 per hour, based on houseworker salaries published by the Bureau of Economic Analysis. In the 2003 Bureau of Labor Statistics time use study, the number of household hours for each of the four groups was multiplied by each group's respective U.S. population to calculate the total number of hours of household work performed: 296 billion. In 2003, the work was valued at $8.23 per hour, based on houseworker wage data from the Bureau of Labor Statistics.

In the 2004 American Time Use Surveys, the data were broken down by sex, status of employment, and age of children in the household. To consolidate the numbers into the four subgroups, we weighted them using demographic data from the U.S. Census Bureau's Current Population Survey. Otherwise, the methodology was the same as that used to calculate household hours worked in 2003. Total hours of household work performed in 2004 were 304 billion, valued at $8.34 per hour. Thus, the GPI estimates the value of housework and parenting at $2.5 trillion in 2004. This represents the largest positive adjustment to personal consumption expenditures. The value of housework and parenting was roughly 33% of personal consumption expenditures in 2004; whereas in 1950 it was 58%. In part, this reflects our increasing reliance on the market to provide services formerly contributed by households.

Item F: Value of Higher Education

There has been considerable debate over whether to include column F at all. Previous editions of the GPI have omitted the cost of higher education, considering it an investment. Other studies have deemed higher education to be consumption, while still others have asserted that the primary value of higher education is a "preparatory" expenditure. While it is clear that the long-term earnings of college graduates are much higher than those without a college degree, we sidestepped the debate over how to address these individual benefits by focusing instead on benefits to society.

Hill et al. (2005) provide an exhaustive list of such benefits, which are both monetary and nonmonetary and in the form of increases in personal and social knowledge, productivity of workers and capital, civic participation, efficiency in the job market, the rates at which people save, the amount and benefit of research and development, charitable giving, and health. Based partially on Moretti (2004), they estimate the effective total value of this social spillover to be $16,000 per year per college-educated worker. We multiplied this value by the number of people twenty-five years and older that had completed at least four years of college as reported in periodic U.S. Census Bureau Current Population Surveys. In 2004, we estimate the annual social benefits of higher education to be nearly $828 billion. This represents the GPI's second largest addition to personal consumption expenditures.

Item G: Value of Volunteer Work

Some of the most important work in America is not done for pay. Such work is not only performed at home, but also in the broader realm of our neighborhoods and communities. Work done here is the nation's informal safety net, the invisible social matrix on which a healthy market economy depends. While it is arguable whether each additional lawyer, broker, or advertising account executive represents a net gain for the nation, there is little question that churches and synagogues, civic associations and neighborhood associations are doing work that is desperately needed. Despite its crucial contribution, however, this work is entirely ignored in the GDP. The GPI begins to correct this omission.

First, we estimated the total number of hours volunteered each year. We relied primarily on three Current Population Surveys conducted by the Bureau of Labor Statistics in 1965, 1974, and 1989 and the American Time Use Surveys from 2003 and 2004. Intermediate years were interpolated. Since the questions asked in each survey were not exactly the same, there are some comparability problems. But the surveys are close enough to provide a workable estimate for the purposes of the GPI.

Second, we applied the Independent Sector estimate of the monetary value of an hour of volunteer time in 2000 (since all GPI figures are reported in 2000 dollar values). That value is $15.68 per hour (Independent Sector 2006). The GPI indicates that the value of volunteer activities in the United States stood at $131 billion in 2004, or $447 per capita. This is significantly higher than the 1950 value of $202 per capita, implying that over the past few decades, Americans have become more generous with their time and that their time is of much greater worth.

GPI Deductions

Item N: Cost of Commuting

Urban sprawl has put more cars on the road, exacerbated traffic congestion, and increased the time Americans must spend getting to and from work. According to the U.S. Department of Transportation (2000), there has been a 66% increase in the number of vehicles per household and significant increases in commute times since 1960. While commuting is for most people an unsatisfying and sometimes frustrating experience, the GDP treats it as a benefit to consumers. The more time and money spent commuting, the more these regrettable activities contribute to the GDP. Moreover, the GDP does not account for the opportunity costs of time spent commuting—time that could be spent freely with family, at leisure, sleeping, or at work.

The GPI corrects this GDP shortcoming by subtracting the cost of commuting. There are two distinct types of costs incurred in commuting. The first is the money spent to pay for the vehicle, or for bus or train fare; the second is the time lost that might have been spent on other, more enjoyable or productive activities. In the GPI accounts, the direct (out-of-pocket) costs of commuting are a function of the portion of non-commercial vehicle miles used in commuting, the cost of user-operated transport, the depreciation of private cars, the portion of commuter miles on public transportation, and the price to use local transportation. Data for these variables were taken from the Statistical Abstract of the United States and the National Income and Product Accounts of the Bureau of Economic Analysis.

The indirect costs of commuting (i.e., the value of time lost) are calculated as the total number of people employed each year, times the estimated annual number of hours per worker spent commuting, times a constant value for time. Because some people regard commuting as part nuisance and part leisure, we assigned a cost of $10.69 per hour (rather than the $13.36 per hour for lost leisure) based on the assumption that 20% of commuting time may be considered leisure. The number of hours per year was derived from survey data on time use by households (Leete-Guy and Schor 1992) coupled with data from the National Household Transportation Survey from 1983, 1990, 1995, and 2001. According to the National Center for Transit Research (2005) at the University of South Florida, the National Household Transportation Survey shows that commuting times have increased by 29.1% since 1983. The estimated cost of commuting in 2004 was $522.61 billion, or $1,778 per capita. Per capita costs have risen by 91% since 1950.

Item P: Cost of Automobile Accidents

The damage and economic loss due to automobile accidents represents a real cost of industrialization and increasing traffic. The GPI uses fatality and injury statistics published in the Statistical Abstract and by the National Center for Statistics and Analysis (2003). Economic losses are based on estimates made by the National Safety Council (2004). The figures cover motor vehicle accidents on and off the road and all injuries, regardless of length of disability, and address lost wages; legal, medical, hospital, and funeral expenses; and insurance administration costs. Property losses are not included because of significant data gaps. The National Safety Council estimates that, on average, each motor vehicle death represents $1,130,000 in economic losses, and each injury $49,700 in 2004 dollars. Economic losses peaked in 1996 at $206.98 billion. In 2004, such losses amounted to $175.18 billion. The National Safety Council attributes this decline to advances in vehicular safety.

Item Q: Cost of Water Pollution

Water is one of the most precious of all environmental assets, yet the accounts of national income provide neither an inventory of the quantity or quality of our water resources nor a cost accounting for any damage to that resource. In the GPI framework, the costs of water pollution arise from damage to water quality and damage from siltation, which reduces the life span of water impoundments or channels. Although this may involve some double counting (insofar as siltation also damages water quality), on the whole the estimates in this column understate damage because of the lack of data on nonpoint sources of pollution.

The cost of damage to water quality begins with a 1972 estimate of $12.0 billion, or $39.7 billion in 2000 dollar values. This is based on the upper range of estimates in three studies of point source damage to recreation, aesthetics, ecology, property values, and water supplies for households and industries (Freeman 1982). Between 1950 and 1972, damage from water pollution is assumed to grow 3% per year, from $20.3 billion to $39.7 billion. Between 1972 and 1992, damages are assumed to increase at a rate corresponding to the per capita increase in spending on water pollution abatement, which grew from $324 in 1972 to $570 in 1992 (Rutledge and Vogan 1994). We assume that per capita expenditures for the abatement of water pollution are roughly correlated with the magnitude of actual damage to water quality. After 1992, data on the control of water pollution is no longer available; thus, pollution and its associated damage to water quality and quantity is assumed to continue growing at 3% per year, from $71.8 billion in 1992 to $102.3 billion in 2004.

Erosion imposes costs in the form of reduced river navigability, siltation of water impoundments, increased flooding, reduced recreational activities, and degraded fisheries. Uri and Lewis (1999) estimated the social cost of soil erosion to be $17.81 billion in 1997. In that year, we estimate total erosion from agriculture and forestry to be 2.02 billion tons. Adjusting for inflation yields a damage estimate of $8.81 per ton of erosion. As sources of siltation, we examined erosion from farming (960 million tons in 2004) and logging (925 million tons in 2004). Tons of cropland erosion comes from the National Resources Inventory, conducted by the Soil Conservation Service in conjunction with Iowa State University from 1982 to 2003. From 1950 to 1981, we estimate that erosion decreased by an average of 1% per year, based on the trend visible in the National Resources Inventory data.

Tons of logging-related erosion comes from an estimate by Hagerman (1992) that forest operations contribute 231 tons of sediment per acre per year. We have assumed Hagerman's estimate applies to clear-cuts, which are 38% of U.S. harvests (U.S. Department of Agriculture 2006). We further assumed that selective cutting contributes only half as much sediment as clear-cuts, or 115.5 tons per acre. To estimate total acreage of forest operations, we relied on 1950–2002 statistics published by Adams et al. (2006). Combining damage to water quality and damage due to siltation we estimate the total cost of water pollution to be $119.72 billion in 2004.

Item T: Loss of Wetlands

Wetlands contain some of the most productive habitat in the world. Yet their value is not represented in economic accounts because the benefits—such as storing, regulating, and purifying water and providing habitat for fish and waterfowl—are generally "public goods," for which there is no overt price. When a farmer drains and fills a marsh, the GDP rises by the increased output of the farm. However, the loss of services from the wetland goes uncounted. The GPI rectifies this by estimating the value of the services that are given up when wetlands are converted to other uses. To do this, we multiply the annual loss of wetlands by $914, the value of an acre of wetland as estimated by a meta-analysis of wetland valuation reviewed by Woodward and Wui (2001). We add this value to an assumed baseline of wetland loss prior to 1950, since we continue to incur the cost of not having these wetlands present to perform essential services, such as water storage and filtration.

The U.S. Fish and Wildlife Service (1997) estimates that 136 million acres of wetlands were filled in North America from the colonial period to 1950. Acreage declined from an original 395 million (including the contiguous lower forty-eight states and Alaska) in the 1780s to about 259 million acres in 1950—a loss amounting to sixty acres an hour for two hundred years. Our estimates of acres of wetland loss are based on the 1997 U.S. Fish and Wildlife Service data published in *Status and Trends of Wetlands in the Conterminous United States*. Their most recent study estimated the loss of wetlands at 462,000 acres per year through 1975, 294,000 acres per year from 1976 to 1984, and 121,000 acres per year in subsequent years. Each of these figures includes 4,000 acres per year lost in Alaska, while the remaining acres were lost in the lower forty-eight states. We extrapolate the acreages of wetlands lost since 1995 by using the rate of change from 1985 to 1995. The GPI estimates the accumulated cost of lost ecological services from filled wetlands in 2004 to be $53.26 billion.

Item U: Loss of Farmland

Loss of either natural or human-built capital generates costs to both present and future generations in the form of lost services from that capital. By destroying farmland, we are losing a vital ecosystem service—a sustainable food supply. Loss of farmland due to urbanization and poor management practices also generates costs in the form of lost scenic, aesthetic, and historic values, increased flooding, deterioration in the quality of water and the fertility of soil, as well as the degradation of wildlife habitat. The GPI accounts for farmland losses associated with urbanization and unsustainable farming practices that reduce long-term productivity.

Obtaining accurate time-series data on the loss of farmland is a surprisingly difficult task. Variations in time periods studied, how farmland is defined, and how acreage is counted are considerable. For this reason, we combined data from a number of sources, including American Farmland Trust, the National Agricultural Statistics Service, U.S. Department of Agriculture's National Agricultural Lands Study, and the Farm Information Center. Using these data sets, we estimate the average annual conversion of prime farmland to urbanization at nearly 400,000 acres per year since 1950.

To put a price tag on this loss, we added the average value ($5,459) from three contingent valuation studies summarized by Ready et al. (1997), which considered lost amenity values to the Costanza et al. (1997) figure of $41.34 per acre for lost ecosystem services. We then multiplied the resulting value ($5,500 in 2000 dollar values) by an index that deflates this value in years before 2000 and inflates it after to account for relative scarcity. By 2004, the GPI accounts assign a cost of $6,203 for every acre of farmland lost to urbanization. The cumulative figure is obtained by multiplying each year's value per acre by the acres lost in that year, then adding it to the previous year's loss. As with wetlands, the reason for tracking cumulative and not marginal losses is the fact that we are still incurring costs of farmland lost in 1950 and every decade thereafter because we are no longer receiving the stream of benefits these lands once conferred (and still could if they are restored). The GPI assumes that the initial pre-1950 loss was roughly $3.31 billion, a figure that grew to $91.19 billion in 2004.

Urbanization removes the productive potential of farmland in a highly visible way. But it may not be as serious in the long run as the deterioration of soil due to poor management. The decline of soil quality over the past forty years has been masked by higher inputs of fertilizer, pesticides, and fuel. In addition, soil depletion is not linear. It may not show up gradually in yield reductions, but rather in a sudden and irreversible decline. Losses of agricultural productivity from erosion have been estimated at $1.3 billion per year, or $2.5 billion in 2000 dollar values (U.S. Department of Agriculture 1985). In 1985, calculations of soil erosion from item Q show 2.9 million tons of cropland erosion in that year, which translates into roughly $0.86 per ton. We assume the cumulative damage prior to 1950 was $16.3 billion, and add to that by multiplying the $0.86 figure by the annual erosion estimated from item Q.

The damage to soil from compaction by heavy machinery in 1980 was estimated at $3.0 billion in 1980 dollars (Sampson 1981), or $5.5 billion in 2000 dollars. We assumed a 3% increase per year in the losses due to compaction prior to and following 1980. The 2004 estimate of the cost of soil compaction is $11.27 billion. The total economic cost of the loss of farmland to urbanization, soil erosion, and soil compaction in the GPI was estimated at $263.86 billion in 2004, having risen steadily from an estimated $25.80 billion in 1950.

Item V: Loss of Primary Forests and Damage from Logging Roads

The ability of native or primary forestland to provide the goods and services required by society is impaired or lost forever when it is converted into tree plantations or cleared to build a road. These goods and services include the forest's ability to control floods; purify air and water; maintain biological, genetic, and functional diversity; provide habitat for sensitive species; produce nontimber forest products; or provide scenic, recreational, and aesthetic values to nearby communities. The GPI measures this loss by assigning a price tag to year-by-year estimates of loss of goods and services within the primary forest, and adding such losses to the cumulative damage from previous years. In particular, we assign costs to the loss of old-growth longleaf pine forests in the southeastern United States; old-growth forests in the Pacific Northwest, Sierras, and southeastern Alaska; and inventoried roadless areas on national forests.

While certainly debatable, we assume relatively little overlap in the damage assigned to loss of roadless areas and old-growth forest largely because roadless areas tend to be located in higher, less productive areas not typically included in inventories of low-elevation, high-productivity old-growth stands. While there are other critical forest types lost in the United States each year, these primary forest types are particularly rich in biological diversity, have been extensively studied, and have reasonable estimates of both extent and value on which GPI accounts can be based. We also incorporate costs associated with logging roads in the national forests, which are a continual source of sedimentation, landslides, fires, and habitat fragmentation.

For longleaf pine, data points for original extent, 1935, 1955, 1985, and 2003, as well as rate of loss in this period, are drawn from Outcalt and Sheffied (1996) and the U.S. Fish and Wildlife Service (2003). Out of an original 60 million acres of longleaf pine forest, only 2.9 million remained in 2004. In the Pacific Northwest, the Forest Service estimates that between 60 and 70% (65% as a midpoint), or 19.57 million acres, of forests within the range of northern spotted owl were in late-successional/old-growth condition during the preindustrial era (U.S. Department of Agriculture 2005). In 1950, we assume that most old growth on private lands had been taken and that national forest boundaries provide a crude proxy for what remained. With this assumption, the 1950 estimate for Pacific Northwest old growth is 15.77 million acres. By 1994, the U.S. Forest service estimated that this figure fell to 7.87 million acres. In terms of annual loss, we assume a rate of loss of 180,000 acres per year between 1950 and 1994. Post-1994 figures are based on losses due to logging and fires reported by the U.S. Department of Agriculture (2005).

Beardsley et al. (1999) estimated the data points in the Sierras for 1945 and 1993. Remaining points were interpolated. In Alaska, we assume that nearly all timber harvests on the Tongass National Forest back to 1950 involved the clearing of old-growth temperate rainforest. Harvest data were taken from spreadsheets provided by the Tongass National Forest.

For inventoried roadless areas, we assume an original extent equivalent to the extent of national forest system lands in the western United States (167 million acres). In 1979, the Forest Service inventoried 62.02 million acres (U.S. Forest Service 1980). In 2000, that figure fell to 58.51 million acres. For intervening years, we incorporated a variety of Forest Service data points on new road construction and multiplied these figures by the amount of roadless area lost per mile of new road construction (26.44 acres per mile).

Taken together, GPI accounts show a cumulative loss of primary forest (the three old-growth forest types discussed above plus roadless areas) equivalent to 183.10 million acres in 2004. To assign a cost to old-growth forest loss we take the figure of $134 per acre (Costanza et al. 1997) for ecosystem services, not including raw materials and climate regulation (since young forests also provide these functions) plus three times that amount for passive use values as estimated by numerous studies, including Vincent et al. (1995). An example of passive use values is the willingness to pay for the preservation of old-growth forest habitat critical to the northern and Mexican spotted owls, a value determined through contingent valuation surveys. To assign a cost to roadless area loss, we incorporate a figure of $56.43 per acre derived from Loomis and Richardson (2000), who studied carbon sequestration, recreation, passive use, and waste treatment benefits of roadless lands. In 2004, the GPI accounts estimated the magnitude of costs associated with the loss of primary forest to be $46.02 billion.

The calculation of environmental damage caused by forest roads on national forest lands is based on the total miles of roads in any given year. A mile of forest road with a sixty-foot right-of-way covers approximately seven acres of land. If impacts (such as noise, edge effects, and runoff) are included, a mile of road affects at least five hundred acres of land. This provides a very rough estimate of the environmental costs because the damage caused by roads depends on many factors, including age, location, slope, the quality of construction, and the frequency of maintenance. Nevertheless, even the best roads cause some continuing ecological disruption by breaking up the landscape, increasing the cumulative effects of soil erosion, disturbing downstream fisheries, and generally increasing the level of human activity.

Estimates of total miles of forest roads are derived from twelve separate Forest Service data points from 1955 to 2004. In the 1995 GPI, we assumed that the cost of damages to forests caused by roads from 1950 to 1959 was $10,000 per mile in 1982 dollars. That figure is here converted to the dollar value in 2000, or $15,939 per mile. From 1960 to 1979, the cost per mile is assumed to decline on a straight-line basis to $7,500 ($11,954 in the 2000 dollar value) per mile due to improvements in road standards. We estimate the cost of ecological damage due to roads at $4.62 billion in 2004. Added together, the GPI accounts show that the loss of primary forest and damage from logging roads amount to $50.64 billion in 2004.

Item X: Carbon Dioxide Emissions Damage

Few scientists dispute the link between carbon dioxide emissions and global warming, or the link between global warming and the increasing incidence and severity of damaging storms, floods, and droughts. As Hurricane Katrina illustrated all too well, this erratic weather is exacting an enormous economic toll each year on our households, infrastructure, and natural capital. Incidence of severe weather, such as Katrina, will escalate insurance payouts, as well as the mounting costs of replacing lost or damaged homes, buildings, livestock, and other household resources. Ironically, these natural disturbances result in a positive feedback loop, whereby the increasing frequency and intensity of severe weather leads to increasing use of natural capital as we rebuild shattered homes and infrastructure in the aftermath. Yet the GDP does nothing to account for the cost of our impacts on the Earth's climate, the increasing costs of cleaning up after severe weather, or the increased depletion of nature's capital. The GPI attempts to address this oversight by assigning costs to carbon emissions.

There are many ongoing studies that attempt to calculate economic damages per ton of carbon emitted into the atmosphere through our burning of fossil fuels. In one recent meta-analysis of 103 separate studies, Tol (2005) found a mean of $93 per metric ton, or $89.57 in 2000 dollar values. Though hotly debated, we adopt this figure as a conservative starting point for incorporating the environmental damage caused by human-caused carbon emissions into GPI accounts.

The GPI relies on carbon emissions data reported by the Oak Ridge National Laboratory in Savannah, Georgia. We assume that only excess emissions are contributing to global warming and deduct the portion of these emissions sequestered by the world's terrestrial and aquatic ecosystems. The Intergovernmental Panel on Climate Change estimates the capacity of the Earth's global carbon sequestration to be 3 billion metric tons of carbon per year (IPCC 2000). Worldwide, anthropogenic (not counting natural emission sources) overshoot of this sequestration capacity began in 1964, and has now risen to 58%, or roughly 4 billion metric tons. In the GPI accounts, we assign costs to a percentage of U.S. emissions that is identical to the percentage of the global overshoot in a given year.[2] We also assume that, due to positive feedback effects, marginal damage increases over time. To account for this, we taper the costs of marginal damage down in retrospect from $89.57 in 2004 to just over zero in 1964—the first year of carbon overshoot. Finally, we assume that marginal damage from carbon emissions is cumulative, so that costs incurred in one year continue to be incurred the next year.

Using this approach, we estimated the environment damage from carbon emissions to be $1.18 trillion in 2004. This is the second largest cost included in the GPI—as it should be. After all, global warming is a phenomenon that threatens hundreds of millions of lives, entire cities, and the planetary economic system like no other threat in human history, and the United States is by far the single greatest source of carbon emissions implicated in that warming.

Item Y: Cost of Ozone Depletion

While annual production of chlorofluorocarbons (CFCs) may have declined dramatically, the cumulative impacts on the depletion of the Earth's ozone layer continue. According to the Climate Prediction Center, "extensive ozone depletion was again observed over Antarctica during the Southern Hemisphere winter-spring of 2005, with widespread total ozone anomalies of forty-five percent or more below the 1979–1986 base period" (National Oceanic and Atmospheric Administration 2006). In September 2005, the area covered by extremely low total ozone values of less than 220 Dobson units (a standard measure of ozone concentration), termed the Antarctic ozone hole, reached maximum size of 25 million square kilometers, with an average size of more than 22 million square kilometers, among the largest sizes of recent years.

There are no definitive studies showing the combined health and ecological consequences of ozone depletion over the next half century. Scientists warn, however, that the loss of ozone could result in increased exposure to harmful solar radiation, which can destroy plants and amphibians, as well as cause cataracts and skin cancer in humans. Given the potentially catastrophic effects on all forms of life, the GPI includes an estimate reflecting our expectation of the economic costs associated with this long-term environmental problem—$45,059 per ton.

The calculation for the cost of ozone depletion involves multiplying the U.S. share of the cumulative world production of chlorofluorocarbons 11, 12, 113, 114, and 115 by $45,059 per ton in 2000 dollar values. To calculate U.S. share, we combined data sets from the Alternative Fluorocarbons Environmental Acceptability Study (www.afeas.org), the Environmental Protection Agency, the United Nations Environmental Programme, and the U.S. Congress. The GPI account estimates the cost of ozone depletion in 2004 at $478.92 billion. Since the production of chlorofluorocarbons in the United States has all but halted, this cost has remained basically unchanged since 1995.

Item AB: The Genuine Progress Indicator

The genuine progress indicator (GPI) starts with personal consumption adjusted for income inequality (item D), adds five items (E through I), subtracts sixteen items (J through Y), and adds or subtracts two items (Z and AA), depending on their sign in a given year. The result is a more honest appraisal—than GDP—of the economic progress of the U.S. economy and the state of its households because it takes into account the benefits of nonmarket activities, education, and services from capital, as well as costs associated with inequality, environmental degradation, and a weakening international position. While incomplete, the GPI demonstrates the value of services derived from real wealth and assets that one could argue are more meaningful in defining the well-being of the nation's households than those tallied by the GDP. The economic exercise embodied in the GPI demonstrates the complexity of accounting for real wealth. If as many economists and statisticians were devoted to this more complete accounting of the state of the economy as they are to GDP, we might be empowered with better information to care for the collective well-being of the nation more prudently.

Item AC: Per Capita GPI

Per capita, the GPI is calculated by dividing the GPI by the U.S. population. Annual figures on the national population are taken from the economic report of the president.

Section 5: Results and Implications

Table 2.1a and b provides a detailed accounting of the GPI in 2004. The starting point is the nearly $7.6 trillion of personal expenditures on consumption reported by the Bureau of Economic Analysis. The GPI disregards this figure to account for financial inequality among citizens, which in 2004 reached its highest level since 1950.

Using the methodology set forth in Anielski and Rowe (1999), we arrive at a weighted figure of just over $6.3 trillion spent on personal

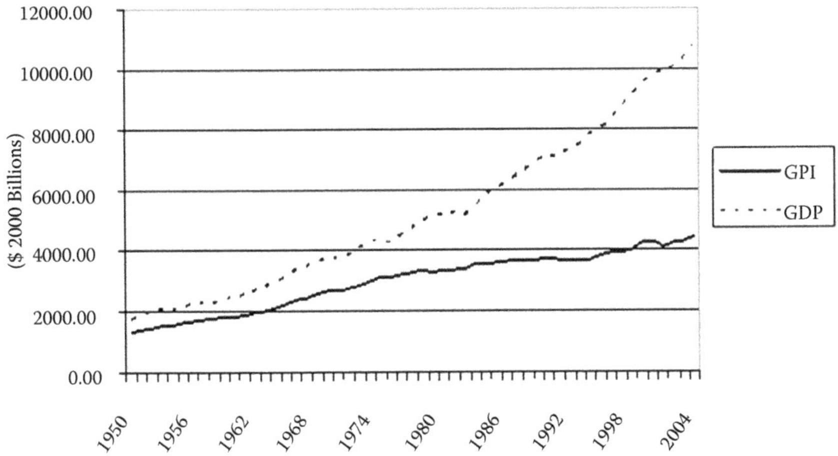

FIGURE 2.1
Real GDP and GPI, 1950–2004.

consumption. To this, we add another $4.7 trillion to account for the benefits of housework, parenting, higher education, volunteer work, and consumer durables, services of streets and highways, and net capital investment. Total positive contributions to the GPI in 2004 amount to roughly $11 trillion. From this, we deduct the environmental and social costs described in Table 2.1b, as well as net foreign borrowing, to arrive at the GPI estimate for 2004: $4.4 trillion, less than half the GDP estimate for that year.

In Figure 2.1, we show GPI and GDP side by side over the 1950–2004 period. Real GPI has increased from $1.31 trillion in 1950 to $4.42 trillion in 2004. This corresponds to an average growth rate of 4% for that time. By comparison, GDP grew steadily from $1.78 trillion in 1950 to $10.76 trillion in 2004, an average annual growth rate of roughly 9%.

These figures mask the effects of increasing population. Thus, it is important to look at both GPI and GDP figures in per capita terms. As shown in Figure 2.2, per capita GPI has barely moved since 1978, remaining near $15,000. Over the 1950–2004 period, GPI grew at an extremely sluggish rate of just 1.33%. In contrast, per capita GDP rose precipitously from $11,672 in 1950 to $36,596 in 2004—an annual growth rate of 3.81%. It is also critical to look at annual growth rates for each year so that important trends within particular time periods are not overshadowed by the full time-series.

Figure 2.3 compares annual per capita GDP and GPI rates of growth using a rolling three-year average to smooth out year-to-year fluctuations. Here, we find a rather striking trend: while GDP growth rates have more

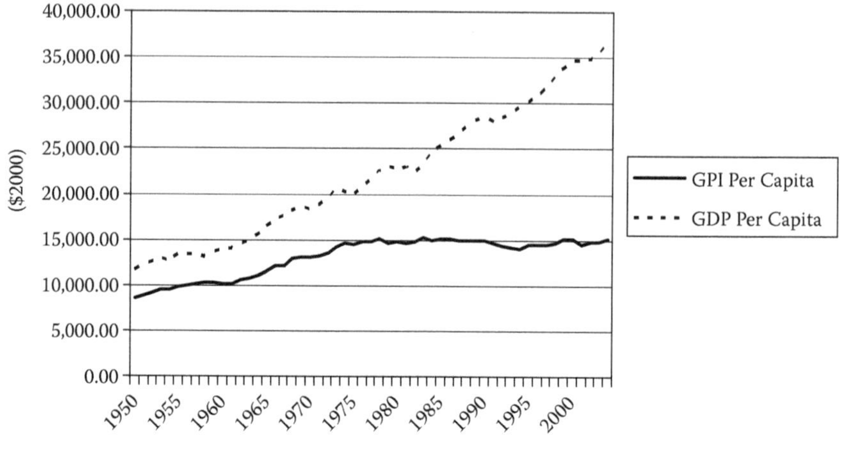

FIGURE 2.2
Real GDP and GPI per capita, 1950–2004.

or less fluctuated within a positive range, the GPI growth rates fall into two distinct periods. In the first period, spanning 1950 to 1980, per capita growth of the GPI more or less matches that of the GDP and was generally positive, ranging as high as 4%.

Beginning in 1980, however, GPI growth rates are commonly negative, bottoming out at 1.64% in 1994. Moreover, the per capita GPI has more or less stagnated since 1978, when it surpassed $15,000 for the first time. Thus, since 1980 or so the marginal benefits associated with growth in expenditures on personal consumption, nonmarket time, and capital services

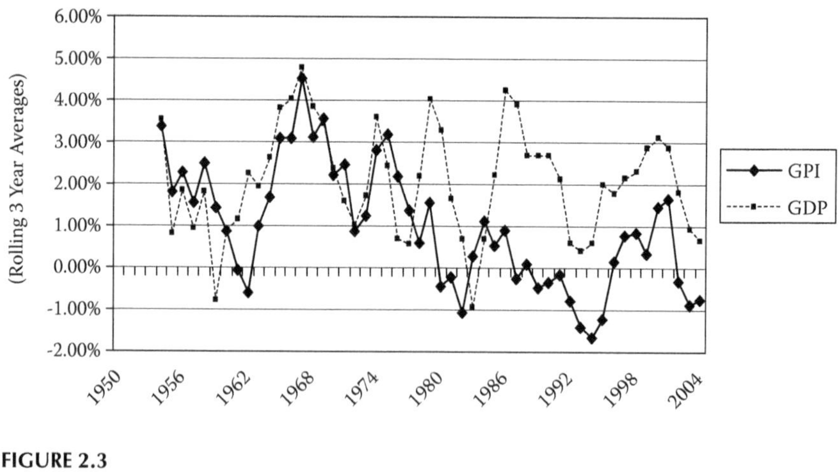

FIGURE 2.3
GDP and GPI per capita growth rates, 1950–2004.

have been offset by the marginal costs associated with income inequality, depletion of natural capital, consumer durable expenditures (i.e., money spent on appliances), defensive expenditures (such as crime prevention), undesirable side effects of growth, and net foreign borrowing. This trend, found in many of the GPI and index of sustainable economic welfare studies completed over the past fifteen years or so, has been put forth as evidence of a threshold effect:

> For every society there seems to be a period in which economic growth brings about an improvement in the quality of life, but only up to a point—the threshold point—beyond which, if there is more economic growth, quality of life may begin to deteriorate. (Max-Neef 1995, 117)

Dietz and Neumayer (2006) argue that the threshold effect found in most GPI and index of sustainable economic welfare studies is probably more an artifact of methodological flaws than a true reflection of the growth and decline of personal welfare. As a case in point, they argue that assumptions made about growth in the costs of depleting nonrenewable resources and incurring long-term environmental damage make the threshold effect all but certain. While their criticisms certainly have merit and warrant closer inspection of the relationship between the threshold effect and actual column-by-column assumptions, we believe this update has at least partially remedied some of those concerns. For instance, in the calculation of long-term environmental damage, we have discarded any assumptions about growth in this damage and, instead, tier damage calculations to actual carbon emissions and the estimated social costs of those emissions. In several other columns, assumed growth rates were replaced by actual data, so it remains unclear to what extent the "hardwired" threshold effect hypothesis of Dietz and Neumayer (2006) still applies.

The growth and relative importance of the contributions to and deductions from the GPI over time is also of interest. Following Lawn (2005), we condensed the items considered by the GPI into several groups. On the contributions side, we left weighted expenditures on personal consumption alone, and grouped items E through I into two categories: nonmarket time (items E, F, and G) and capital services (items H and I). While the absolute magnitude of each has grown steadily, the relative contributions of expenditures on personal consumption and nonmarket time have changed. In 1950, personal expenditures on consumption accounted for 51% of all positive contributions to the GPI. In 2004, that share had risen to 57%. The increasing relevance of expenditures on personal consumption has been accompanied by a corresponding decrease in the relevance of nonmarket time spent on volunteer activities, parenting, and higher education. This share has fallen from 41% in 1950 to 32%

in 2004. As briefly noted in the discussion of item E, this may reflect an increasing reliance on the market to provide services formerly contributed by households (such as home cooking) and a general decrease in our availability to volunteer, extend our formal or informal education, or participate in civic activities.

As for GPI deductions, one significant trend that jumps out dramatically is the growing relevance of costs associated with depletion of and damage to natural capital. This share, which includes loss of wetlands, farmland, and primary forest, depletion of oil reserves, and carbon dioxide emissions and ozone damage, rose from 35% of the GPI deductions in 1950 to 59% in 2004. A large component of this $3.8 trillion cost is the $1.18 trillion in damages associated with carbon dioxide emissions. One reason for this large cost is the assumption the damage is cumulative. In other words, the GPI assumes that we are still incurring the cost of deleterious carbon emissions from 1950 and later.

Dietz and Neumayer (2006) take issue with this assumption, however, and argue, instead, for counting only the marginal—not the cumulative—social cost of carbon emissions. In support of their argument, they point out that most marginal cost figures incorporate the present value of future costs, so tracking cumulative costs, instead of marginal ones, involves double counting.

However, global warming is replete with self-reinforcing feedback loops. For example, melting ice sheets diminish the albedo effect (reflection of the sun's light and warmth back into space), which, in turn, leads to greater oceanic warming. Given the existence of self-reinforcing feedback effects, it would be inaccurate to assume constant marginal costs or somehow neglect the importance of atmospheric thresholds for carbon dioxide beyond which catastrophic effects are more likely.

To their credit, Dietz and Neumayer (2006, 200) suggest increasing the figure for marginal damage over time in recognition of the fact that "the marginal social cost of each tonne of emissions is a positive function of the accumulated stock of carbon in the atmosphere." So something beyond constant marginal cost accounting is appropriate, but it is not clear what that is. Currently, the GPI treats the cost of carbon emissions as cumulative, and increasing over time, but reduces the magnitude of such costs by counting only excess anthropogenic emissions over and above the Earth's ability to sequester those emissions. Due to the ongoing murkiness of exactly how to deal with carbon emissions, we suggest that the methodology presented in this 2006 GPI update be viewed as simply one approach among many potential approaches that should be properly vetted in the years ahead.

Section 6: Using GPI as a Guide to Public Policy

Given the subjective aspects of the GPI and lingering doubts as to its methodological rigor, some have argued that it is irrelevant to public policy debates (Neumayer 1999). For example, Carson and Young (1994, 112) have suggested

> a single, dimension, aggregate measure of sustainable welfare will be of little direct use in guiding, shaping, or choosing among government policies because the factors determining welfare cannot be reduced and combined into a single measure that would command widespread agreement and acceptance.

In Chapter 1, Venetoulis argues that the value-laden character of the GPI "may have skewed the results in a way that has been unfavorable to broader acceptance."

Others, including Daly (1996), point out that using the growth of the GDP as a policy target is a fundamentally flawed approach and that even the "poorest approximation" of personal welfare would do a better job of policy guidance. Anielski (2001, 43) goes quite a bit further by asserting that GPI accounts "provide vital information for holistic and integrated policy decision making, covering virtually every area of government policy." Of course, what information policy makers choose to rely on in making their decisions is often more a function of their political orientations, beliefs, and personal relationships. Therefore, regardless of concerns about the GPI's accuracy and rigor, leaders within government and nongovernmental organizations have used the GPI and its variants as a basis for advocacy.

For example, in Alberta, the Pembina Institute has been publishing GPI accounts since 2001 as a way to persuade the provincial government to adopt a more comprehensive accounting framework that is "capable of assessing the full benefits and full costs of all forms of capital in Alberta—human, social, natural and built."[3] In Nova Scotia, the organization GPI Atlantic reported that the provincial government had created an Office of Health Promotion responsible for all matters relating to the promotion of health, wellness, and addiction, based in part on GPI subaccounts documenting the enormous toll ($3 billion) of largely preventable chronic diseases. As a result, they conclude, "the significance of this cannot be understated: GPI Atlantic is having an impact on public policy."[4] In the San Francisco Bay Area, the quasi-governmental Bay Area Alliance for Sustainable Communities adopted a local variant of the U.S. GPI as a means for tracking progress in achieving the policy objective of a "diversified, sustainable, and competitive economy" (Bay Area Alliance 2004, 12).

The policy relevance of green GDP indicators has also been demonstrated by dozens of peer-reviewed studies. As we previously noted in Section 3, Asheim (2000) found green GDP indicators useful as measures of welfare-equivalent income, sustainable income, and net social profit. Hanley (2000) concludes that green GDP can be used in tandem with more traditional economic indicators to generate useful insights for policy makers seeking to implement broad sustainability goals, such as those included in Agenda 21—a comprehensive sustainability action plan adopted by 178 governments in 1992. More recently, Clarke and Islam (2004) estimated an index of sustainable economic welfare for Thailand that further reinforced the threshold hypothesis and underscored the need for welfare-enhancing interventions by governments of nonindustrialized nations seeking to offset the deleterious effects of pursuing economic growth.

Talberth and Bohara (2006) were among the first to use green GDP time-series data to analyze the impacts of changes in policy on personal welfare by focusing on the effects of greater globalization in trade, which is basically synonymous with "economic openness." Using panel data from eight countries with green GDP accounts and an aggregate production function model (which explains economic growth as a function of a nation's endowment of labor, capital, and variables that affect the productivity of these factors of production), they found a strong negative correlation between openness and green GDP and a strong positive correlation between openness and the gap between traditional and green GDP. The effects, however, were nonlinear, implying that greater globalization is beneficial up to a point. Below, we discuss the results of an independent statistical analysis we performed to update their model using the new U.S. GPI accounts presented here, and extend it to policy variables of interest to the debates over tax cuts and urban sprawl.[5]

Economic Globalization

The debate over the effects of economic globalization has regularly captured headlines since the early 1990s, when the World Trade Organization began its attempts to dramatically increase the pace of liberalizing trade. Empirical studies on the effects of globalization fall into two distinct camps. A number of studies have reported on the beneficial aspects of more global trade regimes, noting, for instance, that the expansion of exports raises the rate of economic growth by way of its impact on total productivity (Dar and Amirkhalkhali 2003). Other studies, however, link greater globalization to deteriorating social and environmental conditions, such as increased inequality of income or greater emissions of greenhouse gases (Baten and Fraunholz 2004; Managi 2004). Of course, what is actually being measured in these studies has a significant bearing on the outcome.

Studies relating globalization to higher rates of economic growth rely almost exclusively on the GDP and related measures, while studies that document the deleterious effects of globalization rely on measures outside the realm of traditional growth models. Thus, Talberth and Bohara (2006) suggest that conducting growth studies using green GDP can help bridge this divide because green GDP is a more accurate measure of personal welfare, which explicitly addresses factors of paramount concern to GDP critics, while maintaining components (e.g., personal expenditures on consumption) that are more consistent with traditional notions of economic growth. Thus, they present a model of growth in green GDP using data sets spanning thirty to fifty years from eight countries: Australia, Austria, Brazil, Italy, the Netherlands, Sweden, the United Kingdom, and the United States. In their model, economic globalization was considered along with measures of human and physical capital typically included in models of economic growth.

In standard economic models, growth is assumed to be a function of changes in a nation's stock of both physical and human capital, as well as other factors that may affect the productivity of these inputs, such as economic globalization (Solow 1956, 1957). In their model, Talberth and Bohara (2006) used changes in the percent of the GDP represented by economic globalization (which is the ratio of trade activity—imports and exports—to GDP), the measure of a nation's supply of physical capital (the "gross fixed capital formation"), and the ratio of the non-working-age members of a population to the working-age members of a population ("age-dependency ratio"). The latter is considered relevant to economic growth because the size of the dependent population may constrain investment in activities that enhance productivity (Holtz-Eakin et al. 2004).

In our independent statistical study, we used our new GPI time-series data to replicate and update the Talberth and Bohara (2006) analysis with respect to the United States. Time-series data for the nation's supply of physical capital (gross fixed capital formation) and the ratio of the non-working-age members of the population to the working-age members of the population (age-dependency ratio) were taken from the World Development Indicators data set. Time-series data for economic globalization were taken from the Penn World Tables.

Validating Talberth and Bohara (2006), our modeling suggests a significant negative, nonlinear correlation between growth in the U.S. GPI and economic globalization, a positive relationship with changes in the nation's supply of physical capital, and a negative relationship with the ratio of the non-working-age members of the population to the working-age members of the population. The results provide some empirical support for the burgeoning literature associating greater globalization with environmental degradation, income inequality, and an increase in economic activity that may be self-canceling from a perspective of personal welfare. Our results also suggest a cautionary approach to the policy of trade liberalization, which is cognizant of the fact that liberalization may be counterproductive past a particular threshold.

Tax Cuts

Tax cuts have been one of the most visible economic policy debates since the Bush administration took office in 2001. The debate has been a bone of contention in both policy and academic circles. In the context of standard growth theory, tax cuts can stimulate long-term economic growth through five main channels, depending on the type and incidence of the particular tax involved. In particular, they can (1) encourage productivity-enhancing investments in the capital stockpile, (2) encourage growth in both the quality and quantity of the labor force, (3) stimulate research and development, (4) steer capital investment to sectors with high productivity, and (5) steer workers toward sectors with high social productivity (Engen and Skinner 1996). Additionally, in the short run, tax cuts can lead to increases in consumer spending.

On the other hand, tax cuts can harm growth if not matched by a commensurate decrease in government spending; otherwise, they will raise deficits and interest rates. If tax cuts disproportionately benefit the wealthy, the resulting "windfall gains" on those with significant assets may undermine incentives for new investments (Gale and Orszag 2005). Tax cuts may also reduce participation in the labor force if the incentive to work more hours at higher pay is more than offset by the incentive to work less and keep income constant (Gale and Orszag 2005). Finally, if tax cuts are matched with decreases in government programs, the socioeconomic benefits of those programs are sacrificed.

Empirical studies relating tax cuts to economic growth are ambiguous. Hashemzadeh and Saubert (2004, 112) assert that "from an historical perspective, there is scarce evidence of a consistent relationship between income taxes and economic growth." They also note that periods of high economic growth in production have correlated quite well with higher taxes. On the other hand, Engen and Skinner (1996) correlate lower taxes with higher growth by predicting a 0.2 to 0.3% boost in economic growth rates associated with a 5% cut in marginal tax rates. Recently, Diamond (2005) predicted that extending the 2001 and 2003 income tax cuts would stimulate investment, employment, and production.

As with the debate over economic globalization, both proponents and opponents of tax cuts have almost exclusively argued their points from a single perspective—economic growth as traditionally defined—rather than from the standpoint of more comprehensive measures of personal welfare like the GPI. Given the empirical and theoretical ambiguity of the debate, and given the paucity of studies relating taxation and personal welfare, a correlation between the GPI and taxes may be a useful exercise, and there are a number of ways the GPI and tax cuts may be related.

If tax cuts exacerbate the inequality of personal incomes, the GPI will fall. If tax cuts cause reductions in beneficial government programs (e.g., farmland conservation, renewable energy, or improvements in water quality), the GPI may also fall. The GPI may also fall because tax cuts often induce an influx of foreign capital (Gale and Orszag 2005). If this capital is used to finance current consumption, the GPI will fall (Anielski and Rowe 1999). On the other hand, if tax cuts boost personal consumption, participation in volunteer work, or educational activities, the GPI could be expected to rise. The GPI may also rise if tax cuts stimulate greater capital investment.

As a preliminary investigation, we modified our statistical analysis of globalization discussed in the last section by adding a tax variable. In particular, we incorporate tax collection time-series data from the National Income and Product Accounts tables published by the Bureau of Economic Analysis. Conceptually, adding a tax collection variable to the framework of the aggregate production function discussed previously is complicated by the fact that growth may induce greater tax collections, and not vice versa. Of course, it is not clear if this concern is as relevant to the GPI as it is to growth of the GDP. In addition, we rely—as with globalization—on growth rates, as suggested by Engen and Skinner (1996), rather than the absolute values of tax collection reflected in the GPI. We also rely on figures for per capita tax collection, not totals. Finally, we slow the tax collection variable so that we are testing the correlation between the change in tax collections of 1963 and 1964 on growth in the GPI between 1964 and 1965. This modification makes intuitive sense if we are testing the proposition that reduced government spending affects personal welfare.

What we found was a strong positive correlation between the change in per capita tax collections and growth of the GPI. This finding is consistent with the historical relationship between higher taxes and high economic growth (as measured by GDP) noted by Hashemzadeh and Saubert (2004). A full investigation of these findings to determine the exact channel by which changes in taxes influence growth in the GPI is beyond the scope of this chapter. Nonetheless, as with globalization, our independent statistical analysis demonstrated the potential use of the GPI data to inform the debate over tax cuts and other adjustments to tax policy.

Growth in Urbanization

In our discussion of globalization and tax cuts, we relied on the framework of the aggregate production function to examine the impacts of policy variables on growth of the GPI. Another potentially useful approach is to explore the impacts of policy variables on the gap between the GDP and the GPI. By looking at the gap, we can simultaneously address changes in economic growth (GDP) and personal welfare (GPI). In particular, during years when the gap is widening, the costs of economic growth are more than offset by the deleterious social and environmental costs of that growth on personal welfare. In years when the gap is closing, positive contributions to the GPI overshadow these costs and economic growth enhances personal welfare.

Talberth and Bohara (2006) modeled the effects of changes in economic globalization, the growth rate of carbon dioxide emissions, and the production of livestock on the gap between GDP and GPI and found each to have a significant, positive influence on the rate at which the gap grew. In our independent statistical analysis, we adapted that model by substituting urban land area per capita (for livestock) as a variable of interest in the debate over urban sprawl. Since urban sprawl is defined as "sprawling, low density, fragmented, automobile-dependent development," we would expect a positive correlation between the extent of sprawl and the amount of urban land area per capita (General Accounting Office 1999, p. 1).

There is little dispute that public policy has a direct influence on the extent of urban sprawl. According to the Environmental Protection Agency, a number of federal urban growth and development programs "intentionally or unintentionally accelerated the spread of low density development and businesses at greater distances from towns and cities."[6] The question is whether urban sprawl enhances or detracts from personal welfare. Despite the negative connotation associated with the term, there are at least two channels by which the GDP–GPI gap can improve with more sprawl, again, defined here as more urban land area per person.

First, it is important to note that urban sprawl is partially driven by the need to accommodate high-volume, low-cost retail "big box" stores, such as Wal-Mart, Home Depot, and Costco, which bring an unprecedented volume and variety of low-cost consumer goods to the public in a single location. If more urban sprawl is associated with a greater abundance and easier access to these low-cost consumer goods, the GPI will likely increase since it is based on personal expenditures related to consumption. But the GDP also includes personal expenditures related to consumption, which means this effect will have little impact on the GDP–GPI gap. However, to the extent that concentrated retail centers free up time otherwise spent shopping in multiple locations, the GDP–GPI gap may improve, provided there is a corresponding increase in time spent volunteering, in educational activities, parenting, or housekeeping, values that are overlooked by the GDP. Indeed, saving personal time has always been one of the most important benefits associated with concentrated retail centers:

> Back in the city, the search for goods, whether pleasurable or not, consumes a great deal of time. Shopping competes with other activities and the geography of retailing has always been driven, in part, by the need to economize on time. Minimizing procurement time underlies the existence of retailers in the first place.... Convenience, one of the most enduring themes of retailing, thus has driven the geographic arrangement of stores through cities and suburbs. (Campbell 1996)

In addition, because the GPI counts the services yielded by public streets and highways, sprawl no doubt enhances this GPI contribution because, by definition, more sprawl means more streets and highways per person. And since these services are not counted in the GDP, sprawl may help close the GDP–GPI gap. On the other hand, the GPI deducts costs associated with longer commutes, auto accidents, carbon emissions, and lost farmland. None of these costs are included in the GDP, and so the gap will widen as these costs escalate. The net effects are thus ambiguous and worth exploring in a more systematic fashion.

There were two key results from our independent statistical analysis. First, we note that our results corroborate earlier findings of Talberth and Bohara (2006) by demonstrating a positive nonlinear relationship between globalization and growth of the GDP–GPI gap, and a positive relationship between changes in the rate of growth of carbon dioxide emissions and the gap. Second, we found a positive relationship between growth in the area of urban land per capita and the gap. This suggests that, on balance, the benefits of urban sprawl are more than offset by the costs associated with traffic congestion, auto accidents, carbon emissions, and lost farmland.

Section 7: Concluding Thoughts and Future Refinements

To reiterate, the genuine progress indicator (GPI) and its variants, such as the index of sustainable economic welfare, were conceived as a way to measure changes in national economic welfare with a single, aggregate index. The GPI considers households as the basic building block of a nation's welfare, and thus begins its accounting exercise with expenditures on personal consumption. To this, the GPI adds benefits associated with activities that enhance personal welfare, such as parenting, housework, volunteering, and higher education, as well as the services that flow from household capital and public infrastructure. The GPI then deducts costs associated with pollution, loss of leisure time, auto accidents, and destruction or outright degradation of natural capital, the accumulation of international debt, and the depletion of resources. The end result is an index that attempts to measure our collective welfare in terms of principles of sustainability drawn from the economic, social, and environmental domains.

In this chapter, we presented an updated methodology for the U.S. GPI and a new set of accounts, which are current through 2004. Our updates are the first significant changes to the GPI methodology since 1998, and incorporated a wealth of new studies and sources of information that have evolved since that time. The accounts suggest that, while the U.S. economy has grown steadily since 1950, our collective welfare may have peaked in the late 1970s and stagnated ever since as the benefits of economic growth have been more and more offset by costs associated with the growing inequality of personal income, loss of time spent on nonmarket activities, and environmental degradation. In addition, the costs of climate change are increasing dramatically, as demonstrated all too well by the disasters in the Gulf of Mexico in the summer of 2005.

While some dispute the GPI's ability to measure sustainable welfare or take issue with its methodological soundness, it has, nonetheless, prompted government and nongovernmental organizations throughout the world to use it as a tool for promoting sustainable policies and for demonstrating the fallacy of relying on the gross domestic product (GDP) as a measure of personal and national welfare. And because the GPI accounts yield historical data going back fifty-four years, it is readily adaptable for use by researchers seeking to test the influence that past changes in policy have on the growth of personal and national welfare. We have also discussed in this chapter how GPI time-series data can be incorporated into standard models of economic growth to inform policy debates involving economic globalization, tax cuts, and urban sprawl.

While future refinements to the GPI will attempt to address some of its outstanding theoretical challenges—such as relating future social–environmental impacts to current personal welfare—the bulk of these new refinements will be focused on developing new sources of information and more precise methodologies of calculation. The GPI accounts would be well served by a new set of valuation studies addressing the personal use of time, depletion of natural capital, and costs associated with environmental degradation, such as air and water pollution, since many of the sources underlying these GPI columns are somewhat dated.

There are a number of changes to calculation methodologies that could be made in response to the latest round of vetting in the literature. For example, Lawn (2005) expresses wholehearted agreement with Neumayer's (2000) critique regarding the methods used to calculate resource depletion, and there is no reason why future GPI iterations could not adopt their recommendations. Taken together, these changes will make the GPI a more accurate, robust, and widely endorsed tool for promoting sustainable development in the decades ahead.

Acknowledgment

The authors thank Noah Slattery for his invaluable research assistance.

Endnotes

1. For a useful compilation of Hurricanes Katrina and Rita damage statistics see http://en.wikipedia.org/wiki/Hurricane_Katrina. For wetland loss associated with the storms see USGS (2006).
2. Of course some would argue that all anthropogenic emissions should be assigned a price since they all contribute to increased carbon dioxide concentrations in the atmosphere. However, it should be noted that even the most ambitious emissions cap proposals acknowledge that there is some acceptable level of warming, implying that at some date (and many cap proposals are in the 1964 range) emissions were not a significant externality.
3. See articles on Alberta, Canada, at http://www.greeneconomics.ca/.
4. See GPI Atlantic newsletter 4, April 2003, available online at http://www.gpiatlantic.org/gpinews/gpinews14.pdf.
5. A link to the complete analysis is provided at http://www.rprogress.org/sustainability_indicators/genuine_progress_indicator.htm.
6. See "About Smart Growth," U.S. EPA, online at http://www.epa.gov/smartgrowth/about_sg.htm#fedrole.

References

Adams, Darius, R. Haynes, and A. Daigneault. 2006. *Estimated timber harvest by United States region and ownership, 1950–2002*. Gen Tech. Rpt. PNW-GTR-659. Portland, Oregon: USDA Forest Service, PNW Research Station.

Anielski, Mark. 2001. *Measuring the sustainability of nations: The genuine progress indicator system of sustainable well being accounts*. Edmonton, Canada: Pembina Institute for Appropriate Development.

Anielski, Mark, and Jonathan Rowe. 1999. *The genuine progress indicator—1998 update*. Oakland, CA: Redefining Progress.

Asheim, Geir B. 2000. Green national accounting: Why and how? *Environment and Development Economics* 5:25048.

Barber, D.A. 2003. The "new" economy. *Tucson Weekly*, January 8.

Baten, J., and U. Fraunholz, 2004. Did partial globalization increase inequality? The case of the Latin American periphery, 1950–2000. *CESifo Economic Studies* 50:45–84.

Bay Area Alliance for Sustainable Communities. 2004. *State of the Bay Area: A regional report*. Oakland, CA: Bay Area Alliance for Sustainable Communities.

Beardsley, Debby, Charles Bolsinger, and Ralph Warbington. 1999. *Old growth forests in the Sierra Nevada: By type in 1945 and 1993 and ownership in 1993.* Research Paper PNW-RP-516. Portland, OR: USDA Forest Service, Pacific Northwest Research Station.

Campbell, John. 1996. *Time to shop: The geography of retailing*. Boston: Federal Reserve Bank of Boston. http://www.bos.frb.org/economic/nerr/rr1996/summer/camp96_3.htm

Carson, Carol S., and Allan H. Young. 1994. The ISEW from a national accounting perspective. In *The green national product: A proposed index of sustainable economic welfare*, ed. Clifford W. Cobb and John B. Cobb. Lanham, MD: University Press of America. pp. 111–134.

Clarke, Matthew, and Sardar M.N. Islam. 2004. Diminishing and negative welfare returns of economic growth: An index of sustainable economic welfare (ISEW) for Thailand. *Ecological Economics* 54:81–93.

Cobb, Clifford W., and John B. Cobb Jr. 1994. *The green national product: A proposed index of sustainable economic welfare.* Lanham, MD: University Press of America.

Cobb, Clifford, Ted Halstead, and Jonathan Rowe. 1995. If the GDP is up, why is America down. *Atlantic Monthly*, October.

Costanza, Robert, Ralph d'Arge, Rudolf de Groot, Stephen Farber, Monica Grasso, Bruce Hannon, Karin Limburg, Shahid Naeem, Robert V. O'Neill, Jose Paruelo, Robert G. Raskin, Paul Sutton, and Marjan van den Belt. 1997. The value of the world's ecosystem services and natural capital. *Nature* 387:253–59.

Daly, Herman. 1996. *Beyond economic growth: The economics of sustainable development*. Boston: Beacon Press.

Daly, Herman, and John B. Cobb. 1989. *For the common good: Redirecting the economy toward community, the environment, and a sustainable future*. Boston: Beacon Press.

Dar, A., and S. Amirkhalkhali. 2003. On the impact of trade openness on growth: Further evidence from OECD countries. *Applied Economics* 35:1761–66.

Diamond, John W. 2005. Dynamic effects of extending the 2001 and 2003 income tax cuts. *International Tax and Public Finance* 12:165–92.

Dietz, Simon, and Eric Neumayer. 2006. Some constructive criticisms of the index of sustainable economic welfare. In *Sustainable development indicators in ecological economics*, ed. Philip Lawn. Cheltenham, UK: Edward Elgar Publishing. pp. 186–208.

Eisner, Robert. 1985. The total incomes system of accounts. *Survey of Current Business*, January.

Engen, Eric, and Jonathan Skinner. 1996. Taxation and economic growth. *National Tax Journal* 49:617–42.

Fisher, I. 1906. *Nature of capital and income*. New York: A.M. Kelly.

Foertsch, Tracy L. 2006. *A victory for taxpayers and the economy. Heritage Foundation WebMemo 082*. Washington, DC: The Heritage Foundation. www.heritage.org/Research/Taxes/wm1082.cfm.

Freeman, Myrick. 1982. *Air and water pollution control: A benefit-cost assessment*. New York: John Wiley & Sons.

Gale, William G., and Peter R. Orszag. 2005. Economic effects of making the 2001 and 2003 tax cuts permanent. *International Tax and Public Finance* 12:193–232.

General Accounting Office. 1999. *Community development: Extent of federal influence on urban sprawl is unclear*. Washington, DC: General Accounting Office.

Hagerman, J.R. 1992. *Upper Little Tennessee River aerial inventory of land uses and non-point pollution sources*. TVA/WR-92/10. Chattanooga, TN: Tennessee Valley Authority, Water Quality Department.

Hanley, Nick. 2000. Macroeconomic measures of sustainability. *Journal of Economic Surveys* 14:1–30.

Hashemzadeh, Nozar, and Wayne Saubert. 2004. The effects of Bush's tax cuts on income distribution and economic growth in the United States. *Problems and Perspectives in Management* 3:111–20.

Hicks, John. 1947. *Value and capital*. 2nd ed. London: Clarendon.

Hill, Kent, Dennis Hoffman, and Tom R. Rex. 2005. *The value of higher education: Individual and societal benefits*. Tempe: Arizona State University, W.P. Carey School of Business.

Holtz-Eakin, D., M. Lovely, and M. Tosun. 2004. Generational conflict, fiscal policy, and economic growth. *Journal of Macroeconomics* 26:1–24.

Independent Sector. 2006. *Independent Sector's giving and volunteering in the United States, annual update*. Washington, DC: Independent Sector.

Intergovernmental Panel on Climate Change (IPCC). 2000. *Land use, land-use change, and forestry*. Geneva: United Nations Environmental Program and the World Meteorological Organization.

Kuznets, Simon. 1934. *National income, 1929–1932*. Senate Document 124, 73d Congress, 2d session, 1934.

Lawn, Philip A. 2003. A theoretical foundation to support the index of sustainable economic welfare (ISEW), genuine progress indicator (GPI), and other related indexes. *Ecological Economics* 44:105–18.

Lawn, Philip. 2005. An assessment of the valuation methods used to calculate the index of sustainable economic welfare (ISEW), genuine progress indicator (GPI), and sustainable net benefit index (SNBI). *Environment, Development, and Sustainability* 2005:185–208.

Leete-Guy, Laura, and Juliet B. Schor. 1992. *The great American time squeeze: Trends in work and leisure, 1969–1989*. Washington, DC: Economic Policy Institute.

Loomis, John, and Robert Richardson. 2000. *Economic values of protecting roadless areas in the United States*. Washington, DC: Wilderness Society and the Heritage Forests Campaign.

Managi, S. 2004. Trade liberalization and the environment: Carbon dioxide for 1960–1999. *Economics Bulletin* 17:1–5.

Max-Neef, M. 1995. Economic growth and quality of life: A threshold hypothesis. *Ecological Economics* 15:115–18.

Moretti, E. 2004. Estimating the social return to higher education: Evidence from longitudinal and repeated cross-sectional data. *Journal of Econometrics* 2004:175–212.

National Center for Statistics and Analysis (NCSA). 2003. *Traffic safety facts, 2003 data*. Washington, DC: National Center for Statistics and Analysis.
National Center for Transit Research (NCTR). 2005. *Public transit in America: Results from the 2001 national household transit survey*. Tampa: National Center for Transit Research, Center for Urban Transportation Research, University of South Florida.
National Oceanic and Atmospheric Administration (NOAA). 2006. *Southern hemisphere winter summary 2005*. Washington, DC: U.S. Department of Commerce, National Oceanic and Atmospheric Administration, Climate Prediction Center. www.cpc.ncep.noaa.gov.
National Safety Council. 2004. *Estimating the costs of unintentional injuries, 2004*. Itasca, IL: National Safety Council.
Neumayer, Eric. 1999. The ISEW—Not an index of sustainable economic welfare. *Social Indicators Research* 48:77–101.
Neumayer, Eric. 2000. On the methodology of ISEW, GPI and related measures: Some constructive suggestions and some doubt on the "threshold" hypothesis. *Ecological Economics* 34:347–61.
Nordhaus, William, and James Tobin. 1972. Is growth obsolete? In *Economic growth*. National Bureau of Economic Research Series 96E. New York: Columbia University Press. p. 17.
Outcalt, Kenneth, and Raymond M. Sheffield. 1996. *The longleaf pine forest: Trends and current conditions*. Resource Bulletin SRS-9. Asheville, NC: Southern Research Station.
Ready, Richard C., Mark C. Berger, and Glenn C. Blomquist. 1997. Measuring amenity benefits from farmland: Hedonic pricing vs. contingent valuation. *Growth and Change* 28:438–58.
Roth, Andrew. 2005. *The bridge to nowhere and the minimum wage*. Washington, DC: The Club for Growth. www.clubforgrowth.org/blog/archives/026323.php.
Rutledge, Gary L., and Christine R. Vogan. 1994. Pollution abatement and control expenditures, 1972–92, estimates for 1992, revised estimates for 1972–91. *Survey of Current Business* 1994:36.
Rymes, T. 1992. *Some theoretical problems in accounting for sustainable consumption*. Carleton Economic Papers 92-02. Ottawa, Ontario, Canada: Carleton University, Department of Economics.
Sampson, R. Neil. 1981. *Farmland or wasteland*? Emmaus, PA: Rodale Press.
Schneider, Friedrich, and Dominik Enste. 2002. Hiding in the shadows: The growth of the underground economy. *Economic Issues* 30, March.
Solow, R.M. 1956. A contribution to the theory of economic growth. *Quarterly Journal of Economics* 70:65–94.
Solow, R.M. 1957. Technical change and the aggregate production function. *Review of Economics and Statistics* 39:312–20.
Stockhammer, E., H. Hochreiter, B. Obermayr, and K. Steiner. 1997. The index of sustainable economic welfare (ISEW) as an alternative to GDP in measuring economic welfare: The results of the Austrian (revised) ISEW calculation 1955–1992. *Ecological Economics* 21:19–34.
Talberth, John, and Alok Bohara. 2006. Economic openness and green GDP. *Ecological Economics* 58:743–58.

Tol, Richard S.J. 2005. The marginal damage costs of carbon dioxide emissions: An assessment of the uncertainties. *Energy Policy* 33:2064–74.

Uri, Noel D., and James A. Lewis. 1999. Agriculture Dynamics and soil erosion in the United States. *Journal of Sustainable Agriculture* 14:63–75.

U.S. Department of Agriculture (USDA). 1985. *Background for 1985 farm legislation*. Agricultural Information Bulletin 486. Washington, DC: USDA Soil Conservation Service.

U.S. Department of Agriculture (USDA). 2005. *Status and trend of late successional and old growth forest*. General Technical Report PNW-GTR-646. Portland, OR: USDA Forest Service, Pacific Northwest Research Station.

U.S. Department of Agriculture (USDA). 2006. *United States forest facts and historical trends*. Washington, DC: USDA Forest Service.

U.S. Department of Transportation (DOT). 2000. *Journey to work trends in the United States and its major metropolitan areas 1960–2000*. Publication FHWA-EP-03-058. Washington, DC: U.S. Department of Transportation, Federal Highway Administration.

U.S. Fish and Wildlife Service (USFWS). 1997. *Status and trends of wetlands in the conterminous United Status and trends of wetland: Projected trends 1985 to 1995*. Washington, DC: U.S. Government Printing Office.

U.S. Fish and Wildlife Service (USFWS). 2003. *Recovery plan for the red-cockaded woodpecker (Picoides borealis)*. 2nd rev. Atlanta, GA: U.S. Fish and Wildlife Service, Southeast Region.

U.S. Forest Service (USFS). 1980. *Final environmental impact statement: Roadless area review and evaluation (RARE II)*. Washington, DC: USDA Forest Service.

U.S. Geological Survey (USGS). 2006. *USGS reports latest land-water changes for southeastern Louisiana*. www.lacoast.gov.

Vincent, James W., Daniel A. Hagen, Patrick Welle, and Kole Swanser. 1995. *Passive use values of public forestlands: A survey of the literature*. St. Paul, MN: University of St. Paul, Department of Economics.

Woodward, Richard T., and Yong-Suhk Wui. 2001. The economic value of wetland services: A meta-analysis. *Ecological Economics* 37:257–70.

3

*Refining the Ecological Footprint**

Jason Venetoulis and John Talberth

CONTENTS

The ecological footprint measures how much of the biosphere's annual regenerative capacity is required to renew the natural resources used by a defined population in a given year. Ecological footprint analysis (EFA) compares the footprint with biocapacity. Despite increasing popularity of EFA, definitional, theoretical, and methodological issues hinder more widespread scientific acceptance and use in policy settings. Of particular concern are how EFA is defined and what it actually measures, exclusion of open oceans and less productive lands from biocapacity accounts,

* An earlier and different version was published in the journal *Environment, Development and Sustainability*, Vol. 10, No. 4, August 2008.

failure to allocate space for other species, use of agricultural productivity potential as the basis for equivalence factors, how the global carbon budget is allocated, and failure to capture unsustainable use of aquatic or terrestrial ecosystems.

This chapter clarifies the definition of EFA and proposes several methodological and theoretical refinements. Our new approach includes the entire surface of the Earth in biocapacity, allocates space for other species, changes the basis of equivalence factors to net primary productivity (NPP), reallocates the carbon budget, and reports carbon sequestration biocapacity. We apply the new approach to footprint accounts for 138 countries and compare our results with output from the standard model. We find humanity's global footprint and ecological overshoot to be substantially greater than the standard model, and suggest the new approach is an important step toward making EFA a more accurate and meaningful sustainability assessment tool.

Definitions and Background

The ecological footprint is a largely heuristic tool that has been widely used in sustainability analyses for over a decade. In addition to its heuristic value, the power of the ecological footprint is thought to lie not only in the absolute numbers it yields, but also in its ability to compare resource demands of different populations in a common currency of global productivity (Ferguson 1999). According to Wackernagel et al. (2002), the ecological footprint is "a measure of how much productive land and water an individual, a city, a country, or humanity requires to produce the resources it consumes and to absorb the waste it generates, using prevailing technology." According to Wackernagel et al. (2005, p. 5), ecological footprint accounts document how much of the annual regenerative capacity of the biosphere is required to renew the resource input of a defined population in a given year. As referenced in this paper, the ecological footprint is a standardized estimate of the Earth's biological carrying capacity required to support humanity's resource use and waste production.[1]

Ecological footprint analysis (EFA) compares the ecological footprint with available biocapacity. It compares biological capacity used against what is available on a renewable basis. Distinguishing between EFA and the footprint is important. By itself, the footprint tells us little about sustainable resource use; it is simply a measure that increases or decreases as our demands on the environment increase or decrease, without telling us whether or not those demands are sustainable. EFA, on the other hand, is purported to measure sustainability. As noted by Wackernagel et al.

(2002), EFA provides a way to "compare renewable natural resource consumption with nature's biologically productive capacity."

When humanity's footprint is smaller than global biocapacity it is considered sustainable. When it is larger, it is reported to be engaging ecological overshoot or running a negative ecological balance. Currently, leading global footprint practitioners estimate the amount of ecological overshoot to be roughly 23% (Loh and Wackernagel 2004). Hence, "when we compare the current Ecological Footprint with the capacity of the Earth's life supporting ecosystems, we must conclude that we no longer live within the sustainable limits of the planet" (Loh and Wackernagel 2004).

In part, because the footprint embodies a vast amount of information in a single quantitative measure and attempts to operationalize well-known concepts of carrying capacity and sustainability, its popularity is burgeoning among sustainability analysts and practitioners in academic, government, nonprofit, education, and business circles. Nonetheless, EFA faces a number of conceptual and practical challenges that may ultimately hinder its broader acceptance and utility as a sustainability assessment tool if it is given serious scrutiny in a policy-making setting. For example, on a conceptual level, Lélé and Norgaard (2005) note that the footprint is "relevant only with respect to a particular choice of ultimate values or variables of interest, or to particular notions of how disparate values should be aggregated." EFA is thought to reflect an implicit optimism in technology to replace lost biocapacity by boosting forest, crop, and fish yields without incurring any long-term ecological costs (van den Bergh and Verbruggen 1999). The larger problem is that despite its evolution within an objective scientific framework, EFA reflects an anthropocentric orientation that excludes significant aspects of sustainability that might be included using a different theoretical approach. To be sure, leading practitioners in the field have promoted the anthropocentric theory of footprint (Wackernagel et al. 2005). EFA, however, need not be so.

Given that EFA purports to measure consumption of resources within the context of sustainability, it seems appropriate to expand the theoretical basis of EFA from people to the rest of nature by refining EFA to signal when our consumption is jeopardizing the long-term viability of nonhuman life. Thus, it is our contention that a study of the human footprint on the environment *should* begin from an ecologically based theoretical perspective. From this perspective, EFA provides a framework for estimating the Earth's biological carrying capacity. Of course, this approach also comes with a value statement that a study of the human footprint on the environment *should* begin from an ecologically based theoretical perspective.

On a practical level, there has been ample criticism of EFA's assumptions, methods, and data. Among our chief concerns include the exclusion of large areas of the Earth from biocapacity, failure to allocate space for the needs of nonhuman species, use of agricultural productivity potential as the basis for equivalence factors used to normalize disparate land types, allocation of the carbon budget, and failure to capture unsustainable use of forests, fisheries, crop land, pasture land, toxins, and fresh water. At a time when the need and demand for sustainability analysis tools appears to be rapidly escalating, it is critical to address these concerns in a rigorous and systematic manner.

We propose refinements to EFA that begin to address these shortcomings and offer a research agenda for further advancements. Our new approach, an ecological footprint approach that employs net primary productivity (EF-NPP), includes the entire surface of the Earth in biocapacity, allocates space for other species, changes the basis of equivalence factors to net primary productivity (NPP), reallocates the carbon budget, and reports carbon sequestration biocapacity. The footprint provides an excellent framework for measuring the extent (area) of humanity's ecological influence within the context of sustainability. The originators, and our colleagues as global leaders in the field, have done much in the way of making continuous valuable advances to EFA (Wackernagel et al. 2005). Our hope is that the new approach (EF-NPP) will contribute to this process by adding to the breadth of the concept and subtleties of the methodology. We suggest that an NPP-based approach may be useful in addressing some of the problems with standard EFA that may thwart broader, deeper acceptance and use of EFA as an objective and meaningful sustainability analysis framework. The remainder of our chapter is organized under five main headings. We underscore the need for changes to EFA in the face of new demands for its use in research and as a tool for evaluating the sustainability of public policies, business practices, and personal lifestyles in the section titled "The Need for Change to Ecological Footprint Analysis." We review pertinent aspects of the standard ecological footprint methodology and discuss critiques in the "The Standard Approach and Core Critiques" section.

Under the heading "An EFA Approach Based on Net Primary Productivity," we introduce the new methodology. We apply the approach, calculate the footprints of 138 countries, and explain differences between our approach and the standard approach using cross-sectional multivariate regression analysis in the section titled "Application of EF-NPP to the Footprint of Nations." Finally, we offer concluding thoughts and discuss future refinements to EFA.

The Need for Change to Ecological Footprint Analysis

While major critiques of EFA have been in existence since the late 1990s, there are three major trends that underscore the need to make significant changes to EFA at this time: (1) greater demands for EFA in academic research, as a tool for evaluating the sustainability of public policies, business practices, and personal lifestyles; (2) growing abundance of modified footprint methods that cloud the distinction between EFA and other kinds of sustainability analyses; and (3) concerted attempts to set international standards for EFA based on an approach that has not substantively addressed major criticisms in the literature.

There are many recent applications of EFA in both natural and social sciences that illustrate the degree to which it has permeated sustainability analyses across disciplines and in an ever-growing variety of research settings. For example, Rosa et al. (2004) explored the effects of two anthropogenic drivers—population and affluence—on a wide variety of global environmental impacts, including greenhouse gas emissions, emissions of ozone-depleting substances, and the ecological footprint. Dias de Oliveira et al. (2005) compared the benefits and environmental impacts of ethanol fuel in Brazil and the United States using the ecological footprint tool. Warren-Rhodes et al. (2003) used EFA to evaluate live reef fish food product consumption in major Asian economies.

EFA has also been applied to social science research. In a cross-national study, Jorgenson (2003) developed "a recursive indirect effects model to estimate the direct, indirect, and total effects of world-system position, domestic inequality, urbanization, and literacy rates" on the size of average ecological footprints. York et al. (2003) tested theoretical propositions derived from human ecology, modernization, and political economy using stochastic formulation to assess what factors were driving the ecological footprint. Hubacek and Jiljum (2003) used EFA to calculate direct and indirect land requirements for the production of exports from fifteen European Union countries to the rest of the world.

Interest in international, national, and local policy applications of EFA is growing as well. Torras (2003) applied EFA to the problem of debt relief, exploring the possibility of compensatory policy that makes pecuniary transfers from rich to poor countries based on ecological footprints and balances. Barrett (2001) demonstrated the value of EFA as a regional planning tool. According to Barrett et al. (2004), "there are a growing number of local authorities that have conducted an ecological footprint for their local authority area and are applying the results." For example, the City of Santa Monica used EFA to gauge the effects of past policies on the footprint. In policy settings, the ecological footprint is increasingly relied upon to model land use scenarios and guide sustainable development.

Corporate leaders are increasingly concerned about ecological footprints, and this concern has led to a profusion of studies exploring ways to reduce the footprint of companies and the entire sector. For example, several studies have addressed ways that the ecological footprint of tourism can be reduced (Cole and Sinclair 2002; Goessling et al. 2002). Using

EFA, Holden and Hoyer (2005) demonstrate that "the environmentally friendly car truly exists." The ecological footprint of aquaculture is the subject of intensive ongoing research (Kautsky et al. 1997; Wolowicz 2005). Recent announcements of footprint reduction programs by Wal-Mart and British Petroleum are clear indications that EFA may play an increasingly important role in corporate sustainability analyses.

Finally, at the personal level, there is rapidly growing interest in personal lifestyle choices that minimize an individual's ecological footprint. According to Seyfang (2003), the ecological footprint is a "touchstone for understanding the obligations of ecological citizens as a justice based account of how we should live." Millions of visitors each year take the popular "footprint quiz" to understand how their consumption choices can be made more sustainable.[2]

With increasing interest and use of EFA have come demands for revised methodologies to guarantee EFA's ongoing usefulness. For example, Aall and Norland (2005) argue for adjustments in the footprint methodology when shifting from a national to a local policy context to ensure the indicator's applicability in local politics and administration. Bastianoni et al. (2004) used the ecological footprint methodology as a basis for developing a "consumer responsibility approach" to assigning responsibility for greenhouse gas emissions. Sonak (2004) has developed a tool called the ecological footprint of tourism to gauge the sustainability of development activities. A diversity of approaches is warranted, given that some applications are sensitive to local context. However, there are studies carried out and reported under the banner of ecological footprint analysis that are something else entirely.[3]

According to the Global Footprint Network (GFN), "the value of the Footprint as a trusted sustainability metric depends not only on the scientific integrity of the methodology but on consistent application of the methodology across analyses."[4] While standards can be a useful way to preserve the integrity of a metric in the face of multiple approaches and misuse, it may be imprudent to set such standards without addressing problems with the most widely used (and for all practical purposes, the informal standard) approach noted in the literature.

The Standard Approach and Core Critiques

The most up-to-date and detailed treatment of the theory, assumptions, and methodology of standard EFA is found in Wackernagel et al. (2005). Other helpful expositions include Ferguson (1999) and Loh and Wackernagel (2004).

Salient Aspects of the Standard Approach

The standard EFA methodology is based largely on Food and Agriculture Organization (FAO) global agricultural ecological zone (GAEZ) suitability indices. For simplicity, we hereafter refer to the standard approach as EF-GAEZ. Aspects of EF-GAEZ most pertinent here involve (1) the choice of land areas included in biocapacity, (2) equivalence factors used to compare and aggregate disparate land types in a common metric, and (3) assumptions about carbon sequestration rates.

EF-GAEZ is acknowledged by Wackernagel and Silverstein (2000)—one of the co-originators and leaders of EFA worldwide—to be derived from a mechanistic worldview that draws heavily from utility theory and an anthropocentric version of environmentalism. EF-GAEZ is "utilitarian resource accounting ... within a positivist's (if not mechanistic) framework" (Wackernagel and Silverstein 2000). Because of this, EF-GAEZ is exclusively focused on human demands and needs, and thus counts biocapacity only in terms of portions of the Earth that can be of direct use by people. As argued by Monfreda et al. (2004), "by focusing the measure on biologically productive areas that provide particular functions to people, rather than on the total amount of photosynthesis generated, the measure becomes sensitive to the quality of the biomass generation and its usefulness for the human economy." Excluded from biocapacity calculations in the standard EFA methodology is 36 billion hectares of land considered too unproductive to support agriculture or aquaculture as well as the outer reaches of the oceans. Under EF-GAEZ, it does not matter if such areas—which include mountains, deserts, tundra, ice sheets, and most of the ocean—are degraded or destroyed because such areas are not counted as areas from which humanity derives sustenance.

Another related aspect of the EF-GAEZ approach is the assumption that all biocapacity is available for sustainable human use, and that none of this capacity is needed to sustain other species that may indirectly contribute to the amount and quality of renewable resources available to future generations. While other species have been given considerable attention by leaders in the field, formal inclusion in the methodology has been limited to one EFA variant (Chambers et al. 2000). In theory, humanity could appropriate 100% of the Earth's biocapacity counted in EF-GAEZ and still have a sustainable footprint.

A second methodological aspect of interest here involves carbon sequestration rates. EF-GAEZ expresses a population's fossil energy footprint in terms of forest hectares needed to sequester carbon emissions after deducting 35% of those emissions sequestered by oceans. The sequestration rate is based on averages from samples of twenty-six forest biomes in 1980 and 1990 and is assumed to be 0.95 metric tonne of carbon (t C) per hectare per year. Thus, for every metric tonne of carbon emitted over and above the amount sequestered by oceans, EF-GAEZ assumes a footprint of 1.05 hectares.

A final concern we raise has to do with the equivalence factors, which allow footprint practitioners to compare ecological values of disparate land types with a common metric. EF-GAEZ's equivalence factors are based on agricultural potential using United Nations' Food and Agricultural Organization (FAO) data. To compare different types of land, EF-GAEZ first creates broad aggregations of land types or "biomes," including crop land, pasture land, forest land, energy land, built space, and marine and inland fisheries, and then derives a common denominator from FAO's global agricultural ecological zone (GAEZ) data set. That common denominator is the GAEZ estimate for the potential of different land types to be converted into agriculturally productive land. Such potential is evaluated with respect to soil, temperature, slope, precipitation, and other factors regardless of whether the land in question is currently covered by trees, grass, or water. The result is a suitability index for land areas.

GAEZ suitability indices provided the basis for equivalence factors that, in turn, are used to estimate biocapacity for each biome. The most recent biocapacity estimates of EF-GAEZ are presented in Table 3.1.

In the first column, land area reported by FAO is presented on a hectares per capita basis. This area is multiplied by the respective equivalence factor to derive the available biocapacity area, reported in global hectares. Note that no biocapacity is explicitly considered available for the absorption of carbon. One reason for not including carbon sequestration (energy) land explicitly may have to do with one of the core assumptions of EF-GAEZ: that land can only serve one purpose. In

TABLE 3.1

EF-GAEZ Biomes, Equivalence Factors, and Biocapacity

Biome	Actual Land Area (ha/per capita)	Equivalence Factor	Biocapacity (global ha/per capita)
Cropland	0.25	2.11	0.53
Pasture land	0.58	0.47	0.27
Forest land	0.64	1.35	0.86
Built space	0.05	2.11	0.10
Marine and inland fisheries	0.39	0.35	0.14
Energy land	0.00	1.35	0.00
	—	—	—
Total	1.90	—	1.90

Source: Derived from Table 2 in Redefining Progress, *The World's Ecological Footprint and Biocapacity* (Oakland, CA: Redefining Progress, 1999) and Loh and Wackernagel (2004).

other words, if a forest produces wood, the assumption is that it cannot also serve other functions, such as carbon sequestration, soil stabilization, or wildlife habitat. Nonetheless, it appears that carbon dioxide absorption factors are internalized in EF-GAEZ calculations, but not made explicit.

Core Critiques

The most comprehensive critiques are summarized by van den Bergh and Verbruggen (1999). It is not our intent to replicate these discussions. Instead, we focus on critiques of greatest relevance to our suggested advances: (1) by excluding significant natural areas from estimates of biocapacity, national footprint accounts fail to recognize the interdependent nature of all ecosystems; (2) equivalence factors, which influence biocapacity estimates, fail to take into account substantive ecological and bioregional disparities; (3) multiuse land is excluded; (4) calculation of the energy footprint is entirely based on forest carbon sequestration rates; and (5) there is no difference drawn between sustainable and unsustainable land use.

A year later, several authors echoed these concerns in an *Ecological Economics* edition devoted entirely to EFA. One commentator went so far as to recommend against the use of EFA as a measuring rod for sustainability, and especially its use as a way of gauging the environmental merits or demerits of activities, projects, and policies (Opschoor 2000). Answering these critiques, an EFA proponent concluded that "despite its limitations the ecological footprint describes a minimum condition for ecological sustainability: footprints must be smaller than the [total] available ecological capacity" (Wackernagel and Silverstein 2000). While both points of view have merit, to this date, important theoretical and methodological weaknesses of EF-GAEZ have yet to be dealt with in a productive way. Instead of taking sides by disregarding or wholly embracing the EF-GAEZ, our intent here is to begin the process of making advances to EFA so as to make it more compatible with ecological realties, more scientifically robust, and more useful as a sustainability evaluation tool.

An EFA Approach Based on Net Primary Productivity

We propose an initial set of changes to the theory and methodology of standard footprinting that respond to some of the basic critiques outlined by van den Bergh and Verburggen (1999). Because net primary productivity (NPP) is critical to a number of assumptions and calculations inherent to the new approach, hereafter, we refer to it as EF-NPP. According to Running et al. (2004, 547), "NPP marks the first visible step of carbon accumulation; it quantifies the conversion of atmospheric CO_2 into plant biomass." Thus, NPP is a rate process that tracks the net flux of carbon from the atmosphere into green plants per day, week, or year. NPP is highly variable year to year and seasonally. For some seasons and biomes NPP may be negative, indicating that plant respiration is greater than the uptake of carbon by plants, as during months when vegetation is stressed by drought conditions or low temperatures. In addition, succession can influence NPP through allocation of fixed carbon to maintenance rather than growth. So even within a single biome type there is a high degree of variability.

NPP provides the basis for maintenance, growth, and reproduction of all consumers and decomposes. Because of this, NPP is also referred to as a measure of the "total food resource" available on the planet (Vitousek et al. 1986). Because human beings appropriate NPP to fuel production and consumption activities, and because these activities, in turn, affect NPP availability in the future, NPP is particularly relevant in sustainability analyses and seems useful as the basis for EFA accounts that attempt to put disparate types of land into a common currency. In fact, it has been suggested that human appropriation of NPP is "a more explicit measure of the intensity of human pressure on ecosystem use than the ecological footprint, which focuses more explicitly on demand" (UNEP 2005). On the other hand, proponents of EFA argue that human appropriation of NPP fails to indicate anything useful about sustainability thresholds, and that equivalence factors based on GAEZ agricultural productivity data are more robust than actual NPP (Haberl et al. 2004; Wackernagel et al. 2005).[5] A comparative study of EFA and NPP was conducted by Haberl et al. (2004). They suggest that EFA and NPP serve different functions—EFA measures society's utilization of biologically productive areas, while human appropriation of NPP maps the intensity of that use. Rather than extending this debate about the relative merits and drawbacks of the two approaches, what we offer here is a methodology that combines the two by integrating NPP into the EFA framework.

NPP can be incorporated into EFA in a number of useful ways. We suggest four primary changes to EF-GAEZ based on NPP: (1) including the entire surface of the Earth in biocapacity, (2) reserving a fraction of NPP for other species, (3) changing assumptions about carbon sequestration rates, and (4) using NPP as the basis for new equivalence factors.

Including the Entire Surface of the Earth in Biocapacity

As previously discussed, EF-GAEZ excludes areas where resources do not appear (in the data set) to be directly utilized for the purpose of human consumption and waste assimilation. This exclusion, however, disregards the role these areas provide in generating global biocapacity or supporting critical ecosystem services that sustain both human and nonhuman life on the planet. To illustrate this point, productive forests at mid-elevations in western North America are ecologically linked to alpine tundra above and deserts below through the hydrological cycle, wildlife migration, and soil movements, yet EF-GAEZ excludes both deserts and tundra from biocapacity because these areas are determined to have no or extremely low potential for agricultural productivity in the FAO's GAEZ assessments.

From an NPP perspective, however, the entire surface of the Earth is relevant. Because most of the Earth's surface participates in the carbon cycle, the first change in the methodology is to include all land and water area on the Earth as part of biocapacity. The proposed change adds about 36 billion hectares of biocapacity not counted in EF-GAEZ, and primarily consists of areas with relatively low levels of NPP compared to tropical forests, pasture lands, or crop land. This change acknowledges the interconnectedness of the biosphere and is offered as a step toward addressing one of the core critiques of EF-GAEZ (van den Bergh and Verbruggen 1999).

Reserving Habitat for Other Species

Our second change is meant to provide a formal accommodation for other species. As noted earlier, EF-GAEZ takes an explicit anthropocentric stance. As a consequence, the portion of the Earth's biocapacity needed to sustain the diversity of nonhuman life is not removed from the realm of sustainable human appropriation. Nor does EF-GAEZ take other species' needs into account in the context of yield factors used to convert any particular nation's stock of crop land, pasture land, or forest land into global hectares. Because of this, EF-GAEZ has failed to capture the world's biological diversity crisis, indicating that lands we use to meet our demands for food, fiber, timber, and fish are all managed sustainably, while all remaining lands are ignored, suggesting that they have no ecological significance.

According to the Global Footprint Network's (GFN) 2004 *Living Planet Report*, global biocapacity for crop land, forest land, pasture land, and fishing grounds is 1.74 global hectares per capita, while humanity's footprint within these biomes is 0.94, implying that humanity can nearly double its consumption of food, fiber, timber, and fish without exceeding ecological limits (Loh and Wackernagel 2004). This lapse is one of the chief drawbacks of EFA noted by leading ecologists. For example, collapsing cod, salmon, and tuna stocks and numerous scientific assessments cast serious doubt on EF-GAEZ's conclusions that fisheries' yields were sustainable from 1960 to 1999 (Pauly and Watson 2001; Jackson et al. 2001).

While other species are not included in the EF-GAEZ approach to biocapacity estimates, by expanding the (ecological and ethical) boundaries of the biological community as the basis of the EF-NPP approach, an initial step is offered here. Conceptually, since NPP is a food source available to all species, it follows that a certain amount must be removed from the realm of human appropriation to meet other species' needs for food and habitat. A recent scientific assessment found that humans presently appropriate approximately 32% of planetary NPP, a "remarkable level of co-option for a species that represents roughly 0.5% of the total heterotroph biomass on Earth" (Rojstaczer et al. 2001; Imhoff et al. 2004). Considering just those areas accessible to humans, Sundquist (2008) found this figure to range between 89 and 96%. Regardless, evidence strongly suggests that by appropriating the lion's share of NPP on the planet, we have endangered vast numbers of other species and contributed to an extinction rate up to one thousand times greater than background levels (Levin and Levin 2002).

Within the EFA framework, addressing other species' needs can be accomplished in several ways. From an NPP standpoint, it would be necessary to convert all spatial measures of footprint and biocapacity into NPP equivalents (i.e., appropriated vs. available NPP), then "reserve" some percentage of average annual NPP within each biome for other species by deducting that amount from biocapacity. The amount of reserved NPP would have to be based on biome- and sub-biome-specific estimates of ecological sustained yield (ESY). It may be possible to make use of well-established relationships between NPP removal and biological diversity losses to develop these ESY benchmarks for crop land, pasture land, marine and inland fisheries, and forests. If the footprint exceeds these ESY thresholds, it would signal that appropriation of NPP had passed a level commensurate with sustainability of ecosystem health. To be most useful, global EFA accounts would have to provide these signals in advance of ecosystem collapse. Daniel Pauly and others at the Sea Around Us project at the University of British Columbia have taken one important step in this direction by converting catch data now reported in tonnes into tonnes of primary productivity required to support that catch (UBC 2008). However, there are formidable hurdles to this overall approach.

For example, there are situations where intensive land uses such as monocropping increase NPP but harm plant and animal diversity. In addition, large-scale disturbances such as stand-replacing fires may significantly decrease NPP in the short term, but result in improved productivity and species diversity over longer timescales. This makes absolute NPP figures a difficult basis for estimating ESY thresholds in an EFA framework.

In lieu of an NPP reservation strategy, EF-NPP takes a habitat-based approach and sets aside a portion of biocapacity in each biome for other species based on recent global hot spot and gap assessments. According to the latest global assessment by Mittermeier et al. (2004), thirty-four areas totaling 2.3% of the world's surface qualify as biodiversity hot spots—severely threatened places with exceptional endemism and in need of immediate conservation attention. These are largely tropical or subtropical forests threatened by human activities. Average NPP within these areas exceeds the global average of 1.77 pedagrams carbon (Pg C) per year by a factor of at least two and up to nine in some areas. By applying equivalence factors to the actual composition data reported in Mittermeier et al. (2004), we estimate that 15.1% of the Earth's biologically productive space would need to be removed from biocapacity accounts to protect these few hot spot areas. Global gap analysis is another form of global conservation needs assessment. Existing gap studies suggest that if approximately 13.4% of the terrestrial land on Earth were protected, 55% of all species that are significantly threatened with extinction would meet targets for survival (Rodrigues et al. 2003). In regions "with high levels of species richness and endemism ... larger percentages of their territory [require protection]" (ibid.).

In lieu of more thorough needs assessments based on combinations of NPP density studies, gap, hot spots, and other, more site-specific approaches, we use the conservative gap estimate as a starting point and deduct 13.4% of each biome from biocapacity. We would suggest that this is a conservative estimate of the amount of aquatic and terrestrial space actually needed to ensure the well-being of nonhuman life. Nonetheless, it is an adjustment that recognizes the critical importance of providing space for other species within the EFA framework to avoid the pitfalls inherent to EF-GAEZ.

Changing Assumptions about Carbon Sequestration

EF-NPP makes two changes with respect to carbon sequestration: (1) reassigning the carbon budget from forests alone to the entire surface of the Earth and (2) changing the assumed rate of carbon sequestration. One of the more problematic aspects of EF-GAEZ is its assumption that land only serves one purpose at a time (van den Bergh and Verbruggen 1999). The most conspicuous manifestation of this problem is in the way EF-GAEZ treats carbon emissions. In particular, EF-GAEZ assigns the biosphere's entire carbon footprint to forests, but reports no corresponding carbon sequestration biocapacity (Table 3.1). This is because forests are already counted in biocapacity for their role in supplying wood products. EF-NPP resolves this quandary by allowing for multiple land uses. The extent to which multiple land uses are operationalized under EF-NPP is limited to the carbon absorption service, i.e., a hectare of forest can now produce paper *and* absorb carbon. We also recognize the carbon sequestration function provided by all other biomes, so we reassign the carbon budget

TABLE 3.2

World Biocapacity Estimates for EF-NPP

Biome	Area (ha/cap)	Area Adjusted for Other Species (–13.4%)	Equivalence Factor	Biocapacity (global ha/cap)
Crop land	0.25	0.22	2.12	0.46
Forest land	0.62	0.54	3.29	1.77
Pasture land	0.57	0.49	2.42	1.20
Built space	0.05	0.04	0.50	0.02
Less productive land	0.87	0.75	1.04	0.78
Marine and inland fisheries	0.38	0.33	2.67	0.87
Open ocean	5.60	4.85	0.48	2.34
Energy land	n/a	n/a	n/a	8.27
Total[a]	8.34	7.22	—	15.71

[a] Figures may not add up exactly due to rounding.

from just forests to include the entire globe and report it as biocapacity in the footprint–biocapacity accounts (Table 3.2).

Of course, this means that EF-NPP biological capacity is nearly twice the area of the planet since every hectare (after deducting 13.4% for other species) is now counted twice—once for its primary function and once for its carbon sequestration function. While this may be difficult to comprehend, we feel that it better reflects the fact that each hectare of land or sea provides multiple ecosystem services.

The second change deals with carbon sequestration rates. For every ton of carbon emitted, EF-GAEZ apportions a 1.05 hectare footprint based on the uptake potential of relatively young forests during two points in time (1980 and 1990), and as noted above, no land is presented as available biocapacity. In addition, EF-GAEZ does not acknowledge carbon sequestered by 36 billion hectares of land and sea excluded from FAO's GAEZ data. In EF-NPP, we consider the net total potential uptake from the entire surface of the Earth as biocapacity for carbon sequestration, and use recent sequestration rates estimated by the Intergovernmental Panel on Climate Change (IPCC). According to IPCC's models, the total combined carbon sequestration of Earth is estimated to be 3.0 gigatons of carbon (Gt C) annually, with oceans sequestering an estimated 2.3 Gt C (IPCC 2004). Net terrestrial uptake is estimated to be 0.7 Gt C annually. Terrestrial uptake potential is actually higher, but land use changes (e.g., deforestation) have decreased this potential. Of the Earth's 51 billion hectares, oceans cover about 36.7 billion and land covers 14.4. By taking a weighted average of net sequestration potential of the land and sea we arrive at the average carbon absorption rate for EF-NPP: 0.06 tonnes of carbon per hectare per year. This means that for every tonne of carbon emitted EF-NPP assigns a footprint of 16.65 hectares, a significant increase over EF-GAEZ. Moreover EF-NPP makes explicit the addition of 8.27 hectares of carbon sequestration land per capita to biocapacity.

Using NPP as the Basis for New Equivalence Factors

The fourth major modification to EF-GAEZ deals with equivalence factors (EQFs), which are the denominators in biocapacity estimates. Recall that EF-GAEZ equivalence factors were based on potential agricultural yields as determined by GAEZ data. Here, we replace the GAEZ suitability indices with NPP. As shown in Table 3.3, EQFs for EF-NPP are the ratio of each biome's NPP per unit of area to the global average. NPP figures for each biome are based on Table 3 from Amthor et al. (1998, 16), which provides area, annual net primary production (NPP), plant carbon content, and soil carbon content for sixteen distinct biomes. These estimates are based on several decades of research after Ajtay et al. (1979), Post et al. (1982), Botkin and Simpson (1990), Gorham (1995), and FAO (1997). According to Amthor et al. (1998, 16), their NPP figures assumed "potential gains and losses are semiquantitative, based on perceived productivity stimulation due to increasing CO_2 and losses that could occur due to warming stimulated increases in decomposition and reduced productivity due to increasing stress."

Clearly, the science of NPP mapping is evolving rapidly and is now reaching the point where continuous satellite-derived mapping is possible (Running et al. 2004). As we discuss, a key future refinement to EFA

TABLE 3.3

Equivalence Factor Calculations for EF-NPP

Biome	Area ($10^{12}m^2$)	Total NPP (Pg/year)	NPP/Area	Equivalence Factor
Crop land	14.80	6.28	0.4243	2.1214
Forest land	36.10	23.76	0.6583	3.2916
Pasture land	29.80	14.41	0.4835	2.4176
Built space	2.00	0.20	0.0997	0.4984
Less productive land	66.10	13.75	0.2080	1.0400
Marine and inland fisheries	21.30	11.38	0.5344	2.6719
Open ocean	343.60	32.95	0.0959	0.4795
Average	—	—	0.20	—
Total	513.70	102.73	—	—

would be to incorporate this real-time information. For now, however, we simply use the Amthor et al. (1998) estimates to demonstrate the technique of using NPP to provide a more accurate measure of relative ecological productivity across biomes than the GAEZ data. To illustrate how EQFs were derived, consider the 2.12 EQF for crop land. Table 3.3 shows that global NPP for crop land is 6.3 (Pg C) over an area of 14.8×10^{12} square meters, or 0.43 Pg C per square meter. The crop land EQF of 2.12 is simply 0.43 Pg C divided by the global average NPP figure of 0.20 Pg C. Thus, EQFs for EF-NPP represent the ratio of productivity of one land type to the average, where productivity is measured in NPP.

Table 3.3 also displays EQFs for less productive lands and open oceans—areas excluded by EF-GAEZ, and the change in the relative values of each biome. For example, EF-GAEZ had built space as more biologically productive than forest land, pasture land, and marine and inland fisheries. EF-NPP shows the converse. Because of this, we suggest that EF-NPP is more closely aligned with basic scientific understanding of the relative ecological value of different land types. As a related point, we suggest that EF-NPP better captures the ecological impacts of built space. Recall from Table 3.1 that EF-GAEZ assigns identical EQFs to crop land and built space because of the underlying assumption that all built space is displaced crop land (Wackernagel et al. 2005). In contrast, EF-NPP captures variability in the impacts of built space by deducting from future biocapacity global hectares that are more closely aligned with the actual land type being lost, regardless of whether such lands are crop lands, pasture lands, forests, or desert.

All of the changes just presented affect biocapacity, which rises from about 1.9 global hectares (gha) per capita under EF-GAEZ to 15.71 gha under EF-NPP. Table 3.2 provides a breakdown of EF-NPP's biocapacity estimates for crop land, pasture land, forest land, marine and inland waters, open oceans, less productive land, built space, and energy land. Compared with EF-GAEZ, energy land is the greatest addition (8.27 gha per capita) since EF-GAEZ assigns no biocapacity to this function. EF-NPP also adds 3.11 gha per capita to biocapacity for less productive lands and open oceans, whereas EF-GAEZ assigns none.

Changes to the EFA Template

All of the changes to the EFA template were incorporated into the basic EF-GAEZ Excel-based template used to create global footprint accounts. The template finds its origin in Rees and Wackernagel (1994). In subsequent years, the EF-GAEZ template has been refined at two U.S.-based nongovernmental organizations—Redefining Progress and the Global Footprint Network (GFN), both in Oakland, California. Both organizations are informal leaders in producing the global and national footprint accounts based on the standard (FAO) approach. Both have also published the international footprint accounts with World Wildlife Federation International. Slight variations in the EF-GAEZ methodology incorporated in the template used by RP and GFN have developed over the last several years. However, both approaches remain fundamentally the same and show nearly identical results.

Land use, production, and consumption data primarily from the FAO Statistical Database, International Energy Agency, and IPCC form the primary inputs into that template. The template contains equivalence factors and algorithms for estimating yield factors based on these data. As previously discussed, EQFs form the basis for biocapacity calculations. Yield factors form the basis for footprint calculations. For example, at the global level the unadjusted (i.e., before conversion to global hectares) footprint calculation for beans can be expressed as the ratio of crop land area devoted to bean production divided by its yield factor. When that land area is converted into global hectares the size of the footprint changes accordingly. For example, if the EQF goes down one year, so does the footprint because less biocapacity is assumed to be utilized. A more thorough discussion of the data and template operations underlying EF-GAEZ can be found in Wackernagel et al. (2005).

Application of EF-NPP to the Footprint of Nations

We compare global biocapacity and footprint accounts for 2001, as well as global trends between 1961 and 2001, using EF-NPP and EF-GAEZ. We also describe some key differences at the country level and use multivariate regression analysis to systematically evaluate differences in ecological balances under the two approaches.

Global 2001 Snapshot

Use of EF-NPP results in significant changes to global footprint accounts. Table 3.4 shows global per capita biocapacity, footprint, and ecological balances (biocapacity–footprint) for both EF-NPP and EF-GAEZ. EF-NPP shows negative ecological balances (overshoot) in 2001 in four biomes: crop land, marine and inland fisheries, built space, and energy land. EF-GAEZ shows overshoot for energy land alone, a finding corroborated by GFN's 2004 *Living Planet Report*.[6]

Changes made to EQFs and biocapacity as well as deductions made for other species help explain why EF-NPP shows negative ecological balances for more biomes than EF-GAEZ. On a per capita basis, global crop land is determined to have a higher EQF using EF-NPP (2.12 vs. 2.11). However,

TABLE 3.4

Global Footprint Accounts: EF-NPP and EF-GAEZ

	Biocapacity		Footprint		Ecological Balance	
Biome	**EF-NPP**	**EF-GAEZ**	**EF-NPP**	**EF-GAEZ**	**EF-NPP**	**EF-GAEZ**
Crop land	0.461	0.527	0.521	0.527	–0.060	0.000
Forest land	1.775	0.833	0.464	0.189	1.311	0.644
Pasture land	1.197	0.267	0.470	0.091	0.726	0.176
Built space	0.020	0.100	0.046	0.100	–0.026	0.000
Less productive land	0.779	—	0.000	—	0.779	—
Marine and inland fisheries	0.873	0.132	1.045	0.138	–0.173	–0.006
Open ocean	2.337	—	0.000	—	2.337	—
Energy land	8.265	—	19.357	1.142	–11.092	–1.142
Total	15.707	1.859	21.903	2.187	–6.197	–0.328

Note: All figures in global hectares per capita, 2001 data.

deductions made for other species offset this effect so that biocapacity is significantly lower relative to EF-GAEZ (0.461 gha vs. 0.527). The net effect is an overshoot of 0.060 gha per capita, while EF-GAEZ results indicate a zero ecological balance. For marine and inland fisheries, EF-NPP shows a significantly greater EQF (2.67 vs. 0.35), biocapacity (0.873 gha per capita vs. 0.132), and footprint (1.045 gha per capita vs. 0.138). The net effect is an overshoot of 0.173 gha per capita. EF-GAEZ results indicate a zero ecological balance. For built space, EF-NPP has a significantly lower EQF (0.50 vs. 2.11), biocapacity (0.02 gha per capita vs. 0.1), and footprint (0.046 gha per capita vs. 0.1), but overall shows a net overshoot of 0.026 gha per capita. Again, EF-GAEZ shows a zero ecological balance.

Energy land is by far the largest footprint category under both EF-NPP and EF-GAEZ, but the EF-NPP footprint is nearly seventeen times greater. This is largely a function of including the entire Earth in the biocapacity-sequestration estimates, thereby reducing carbon sequestration rates from 0.95 t C/ha to 0.06 t C/ha. EF-NPP also adds 8.27 gha of energy land per capita to biocapacity. The result is an overshoot of 11.1 gha. EF-GAEZ shows an overshoot of 1.14 gha.

Both approaches show positive ecological balances for pasture land and forest land, with EF-NPP showing greater balances due to its higher EQFs for these biomes. Unfortunately, neither approach accounts for desertification of grasslands from overgrazing, salinization of crop land, or loss of forests from unsustainable logging and land conversions. For example, neither approach captures the 1.5 million hectares of crop land lost to salinization each year (Wood et al. 2000), or the 140 million hectare decline in forest area covering the Earth from 1961 to 2001 (FAOSTAT 2005). While such declines may be reflected in lost biocapacity in the future, they are not reflected in negative ecological balances in the present because EFA data are not real time. Future EFA iterations may benefit from remotely sensed data increasingly available from NASA data centers that capture changes in land cover at extremely fine temporal scales on a seasonal and annual basis. Incorporation of these data into EFA is an exciting possibility in the years ahead. In the meantime, EFA does not reflect lost biocapacity from these land use changes and, instead, assumes that biocapacity "breaks even" as a result of higher yields obtained from smaller areas. Of course, higher yields are often associated with ecologically tenuous practices, such as when native forests are converted into tree plantations or plants and animals are injected with growth hormones. EFA's general failure to differentiate between ecologically sustainable and unsustainable practices remains one of its major shortcomings. We lay out an NPP-based research agenda for addressing these and other concerns in our conclusion.

As previously noted, EF-NPP allocates 3.11 gha per capita of less productive and ocean space for biocapacity, while EF-GAEZ fails to assign any. At this time, EF-NPP does not calculate footprints within these biomes. Future advances in EFA data and methods may make this possible. For now, these areas remain unqualified additions to biocapacity.

In the final tally, EF-NPP shows 15.71 gha of total biocapacity, most of which is carbon sequestration land. EF-GAEZ shows about 1.86 gha per capita. The respective footprints are 21.90 and 2.19 gha per capita. Both approaches show a negative ecological balance, or overshoot: EF-NPP, –6.20 gha per capita; EF-GAEZ, –0.33. On a per planet basis, if you will, EF-GAEZ shows a footprint of 1.18 planets. That is, humanity's ecological footprint would require biocapacity the size of another planet that is 18% the size of Earth (at average biocapacity levels) to be sustainable. EF-NPP shows a footprint of 1.39 planets, a 21% increase over EF-GAEZ.

Global Trends over Time

Figure 3.1 illustrates trends in biocapacity and ecological footprints under both approaches between 1961 and 2001. Both EF-GAEZ and EF-NPP show no significant change in biocapacity through the period. Global ecological footprints have risen steadily under both approaches, but more steeply under EF-NPP. Both illustrate that ecological overshoot began in the late 1970s. Thereafter, overshoot has increased to about 18% with EF-GAEZ and 39% with EF-NPP.

On a per capita basis EF-NPP and EF-GAEZ footprints diverge to a considerable extent. Figure 3.2 shows EF-GAEZ and EF-NPP biocapacity and footprints per capita from 1961 to 2001. With EF-GAEZ, there is a rise in the footprint (from 2.61 to 2.79) between 1961 and 1973, then a fairly steady decline through 2001 (from 2.79 to 2.19). This could be due to several

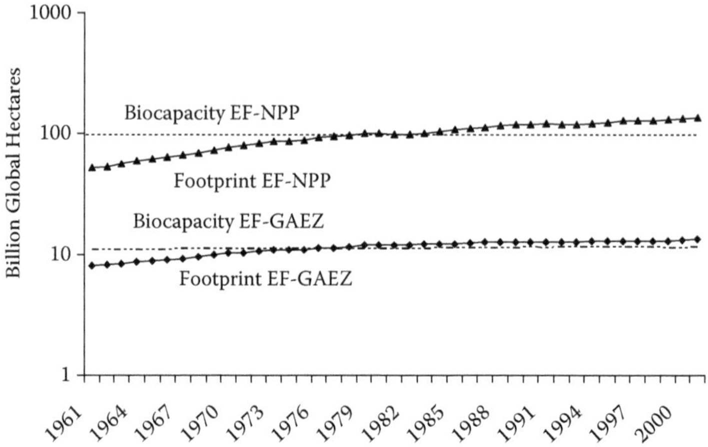

FIGURE 3.1
Ecological footprint and biocapacity, 1961–2001.

factors, including rising population or increases in yield factors. EF-NPP shows per capita footprint increasing over the first twenty years and then becoming fairly stable within the range of 21 to 23 gha per capita thereafter. Under both approaches biocapacity declines by roughly half.

Table 3.5a to d provides EF-NPP footprint, biocapacity, and ecological balance estimates for 138 countries based on 2001 data and indicates differences from EF-GAEZ. Regionally, Africa, Asia Pacific, Latin America,

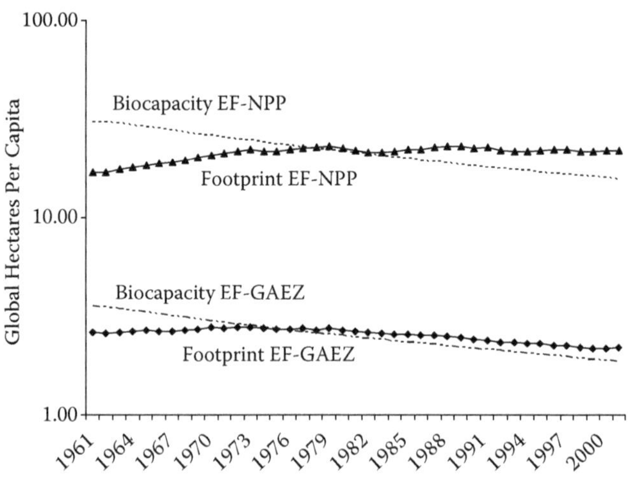

FIGURE 3.2
Ecological footprint and biocapacity per capita, 1961–2001.

and the Caribbean fare better. As compared with EF-GAEZ, these regions show significant gains in their ecological balances. In these regions, the biocapacity of the countries tends to exceed their ecological footprint. In contrast, the Middle Eastern, Central Asian, North American, and European regions tend to show lower ecological balances. Here, EF-NPP shows that most countries are exceeding their biological capacities by a significantly greater margin than what EF-GAEZ suggests. On a country-by-country basis, seventy-five countries showed greater ecological balances, while sixty-three showed lower balances. Mongolia's ecological balance gained the most, rising from 25.17 gha per capita under EF-GAEZ to 163.12 under EF-NPP—a gain of 137.95. The United Arab Emirates saw the most precipitous decline in ecological balance, from –13.77 gha per capita (EF-GAEZ) to –213.43 (EF-NPP)—a drop of 199.66.

Allow us to offer a word of caution concerning these results. The size of the ecological overshoot is largely due to the heavy weight carbon emissions are given in the basic EFA model. While anthropogenic carbon emissions are contributing to a very troubling ecological problem—global climate change—the predominance of the carbon footprint over all other environmental concerns could be construed as presumptuous, and not scientifically defensible. As such, biome-specific accounts may have just as much meaning as the global tally, especially if credible sustainability criteria can be developed.

Cross-Sectional Regression Analysis

To systematically account for ecological balance differences between EF-GAEZ and EF-NPP and to shed light on variables relevant to models that may utilize EF-NPP data in the future, we ran a multivariate cross-sectional regression on the 2001 data. Given the methodological changes embodied in EF-NPP, we can expect that ecological balances would tend to fall in nations that use relatively less energy, have larger land masses, and have greater shares of their biological capacity in pasture, forest, or marine and inland fisheries and tend to rise in smaller nations with relatively higher energy use and greater shares of biological capacity devoted to built space. This is because EF-NPP drastically reduced carbon sequestration rates (which boosts the energy footprint), made significant increases to EQFs for pasture land, forest land, and marine and inland fisheries (which increases biological capacity, especially for large countries), and significantly decreased the EQF for built space (which decreases biological capacity in highly urbanized countries).

To capture energy footprint effects, the regression model uses total gross domestic product (gdp) as a proxy for the absolute level of energy consumption and a nation's energy footprint share (enfpshr) under EF-GAEZ.[7] To capture changes in biocapacity, a nation's EF-GAEZ share

TABLE 3.5A

Ecological Footprint per Capita in Global Hectares (EF-NPP Method) (Africa)

World	Crop Land 0.52	Pasture 0.47	Forests 0.46	Fisheries 1.05	Built Space 0.05	Energy 19.36	Total FP 21.91	Biological Capacity 15.71	Ecological Balance (NPP) −6.20	Ecological Balance (GAEZ) −0.33	Difference −5.87
Africa	0.45	0.80	0.38	0.73	0.04	5.07	7.48	27.51	20.03	2.05	17.98
Algeria	0.64	0.49	0.18	0.13	0.03	13.73	15.21	20.11	4.91	−1.09	6.00
Angola	0.32	0.38	0.22	0.79	0.02	4.68	6.41	44.71	38.30	7.74	30.56
Benin	0.41	0.29	0.49	0.55	0.04	1.28	3.06	10.24	7.18	0.19	6.99
Botswana	0.36	1.91	0.31	0.20	0.02	13.62	16.42	63.61	47.19	2.37	44.82
Burkina Faso	0.57	1.31	0.50	0.11	0.07	0.49	3.05	12.46	9.42	0.44	8.98
Burundi	0.29	0.14	0.62	0.19	0.03	0.23	1.50	7.36	5.86	−0.03	5.89
Cameroon	0.62	0.74	0.37	0.80	0.05	1.85	4.43	18.66	14.23	3.14	11.09
Central African Republic	0.63	1.89	0.45	0.49	0.04	0.69	4.19	47.40	43.21	6.12	37.09
Chad	0.47	1.74	0.46	1.26	0.04	0.24	4.21	32.96	28.75	0.60	28.15
Congo	0.22	0.12	0.31	1.61	0.05	1.15	3.46	55.90	52.44	13.11	39.33
Cote Divoire	0.48	0.20	0.43	0.50	0.06	2.17	3.84	14.52	10.68	1.57	9.11
Egypt	0.48	0.13	0.21	1.14	0.19	10.15	12.30	9.25	−3.05	−0.93	−2.12
Eritrea	0.25	0.84	0.28	0.30	0.04	1.08	2.79	13.63	10.83	0.60	10.24
Ethiopia	0.26	0.14	0.68	0.03	0.04	0.42	1.56	8.53	6.97	−0.24	7.22
Gabon	0.67	0.16	0.24	2.54	0.03	20.43	24.06	120.20	96.15	31.22	64.92
Gambia	0.54	0.73	0.31	0.89	0.04	1.28	3.78	10.64	6.86	0.08	6.78
Ghana	0.42	0.12	0.54	0.87	0.05	1.24	3.23	10.78	7.54	0.39	7.16
Guinea	0.31	0.84	0.73	0.68	0.05	0.90	3.51	21.10	17.59	2.17	15.42
Guinea-Bissau	0.42	1.08	0.32	0.62	0.02	1.20	3.65	28.60	24.95	3.13	21.82

Kenya	0.22	1.51	0.37	0.04	0.04	1.69	3.87	12.70	8.83	0.74	8.09
Lesotho	0.25	1.25	0.45	0.00	0.02	0.75	2.72	12.41	9.69	0.57	9.11
Liberia	0.23	0.06	0.72	0.56	0.04	0.59	2.20	20.39	18.19	2.88	15.31
Libya	0.89	0.37	0.12	0.57	0.03	37.51	39.50	71.09	31.60	–2.20	33.79
Madagascar	0.28	1.54	0.28	0.59	0.04	0.73	3.46	23.85	20.39	2.08	18.30
Malawi	0.29	0.08	0.25	0.48	0.05	0.71	1.86	8.50	6.64	0.24	6.40
Mauritania	0.37	4.32	0.24	0.46	0.04	5.84	11.28	79.77	68.49	0.81	67.67
Mauritius	0.57	0.40	0.23	2.01	0.08	19.02	22.31	58.41	36.11	–0.93	37.04
Morocco	0.51	0.32	0.08	0.56	0.03	4.73	6.22	10.65	4.43	–0.14	4.56
Mozambique	0.23	0.10	0.47	0.18	0.02	0.48	1.49	23.99	22.50	2.83	19.68
Namibia	0.99	1.93	0.00	3.40	0.07	3.28	9.67	106.96	97.30	3.18	94.11
Niger	0.70	0.34	0.16	0.14	0.04	0.54	1.92	26.84	24.91	–0.02	24.93
Nigeria	0.56	0.24	0.34	0.63	0.05	4.02	5.84	9.04	3.20	–0.03	3.23
Rwanda	0.33	0.26	0.48	0.11	0.04	0.44	1.65	7.13	5.48	0.01	5.47
Senegal	0.44	1.17	0.37	1.52	0.04	2.43	5.98	14.46	8.48	0.66	7.82
Sierra Leone	0.25	0.25	0.56	1.12	0.04	1.05	3.27	14.15	10.87	0.79	10.08
South Africa	0.65	1.62	0.60	0.51	0.06	37.19	40.62	20.80	–19.81	–0.32	–19.50
Sudan	0.45	1.68	0.33	0.18	0.04	1.56	4.23	18.99	14.76	1.04	13.71
Tanzania	0.24	1.25	0.34	0.96	0.05	0.55	3.39	15.81	12.42	1.34	11.08
Togo	0.59	0.25	0.60	0.49	0.03	1.87	3.83	9.05	5.23	–0.13	5.35
Tunizia	0.84	0.20	0.26	0.86	0.04	8.14	10.35	10.70	0.35	–0.66	1.01
Uganda	0.51	0.35	0.83	1.62	0.05	0.19	3.55	10.00	6.44	0.05	6.39
Zambia	0.52	0.53	0.41	0.77	0.03	0.63	2.89	30.40	27.51	3.77	23.74
Zimbabwe	0.28	1.11	0.39	0.17	0.02	7.06	9.03	16.35	7.32	1.19	6.13

TABLE 3.5B

Ecological Footprint per Capita in Global Hectares (EF-NPP Method) (Middle East, Central Asia, and Asia Pacific)

World	Crop Land 0.52	Pasture 0.47	Forests 0.46	Fisheries 1.05	Built Space 0.05	Energy 19.36	Total FP 21.91	Biological Capacity 15.71	Ecological Balance (NPP) –6.20	Ecological Balance (GAEZ) –0.33	Difference –5.87
Middle East and Central Asia	0.58	0.55	0.17	0.61	0.05	39.65	41.61	13.55	–28.06	–1.75	–26.32
Armenia	0.38	0.70	0.03	0.05	0.03	4.01	5.20	7.55	2.36	–0.36	2.72
Azerbeijan	0.52	0.96	0.08	0.04	0.03	18.98	20.60	11.64	–8.96	–0.80	–8.17
Georgia	0.41	1.11	0.01	0.04	0.02	6.01	7.60	9.36	1.76	–0.12	1.87
Iran	0.51	0.64	0.04	0.48	0.05	23.65	25.36	12.45	–12.91	–1.04	–11.87
Israel	0.83	0.41	0.50	2.29	0.09	52.30	56.42	6.55	–49.87	–4.14	–45.73
Jordan	0.53	0.28	0.18	0.53	0.06	14.06	15.64	9.06	–6.58	–1.30	–5.29
Kazakhstan	0.71	0.97	0.06	0.04	0.02	33.87	35.66	34.61	–1.05	–0.11	–0.94
Kuwait	0.54	0.35	0.27	0.63	0.15	152.98	154.91	8.41	–146.50	–9.43	–137.06
Kyrgyzstan	0.56	0.85	0.03	0.02	0.03	8.70	10.19	10.19	0.00	8.83	–8.83
Lebanon	0.65	0.24	0.36	0.33	0.07	22.24	23.90	6.33	–17.57	–1.96	–15.62
Saudi Arabia	0.85	0.29	0.22	0.75	0.09	65.89	68.10	23.19	–44.91	–3.31	–41.60
Syria	0.57	0.21	0.06	0.19	0.06	17.27	18.36	7.98	–10.39	–0.85	–9.54
Tajikistan	0.27	0.31	0.02	0.00	0.02	4.02	4.64	9.65	5.01	–0.29	5.30
Turkey	0.82	0.53	0.28	0.56	0.06	14.01	16.25	9.08	–7.17	–0.77	–6.40
Turkmenistan	0.60	0.76	0.01	0.04	0.03	26.42	27.86	25.40	–2.46	–0.59	–1.88
United Arab Emirates	1.18	0.42	0.83	4.22	0.08	226.13	232.86	19.43	–213.43	–13.76	–199.66
Uzbekistan	0.31	0.42	0.01	0.02	0.04	19.78	20.58	20.58	0.00	–1.10	1.10
Yemen	0.28	0.40	0.03	0.68	0.04	3.43	4.87	12.45	7.58	–0.33	7.91
Asia Pacific	0.53	1.09	0.51	1.69	0.44	15.55	19.42	29.97	10.55	2.15	8.40

Australia	1.85	1.84	1.57	2.10	0.05	71.63	79.05	110.21	31.16	4.36	26.80
Bangladesh	0.23	0.03	0.10	0.68	0.03	1.27	2.33	6.50	4.17	–0.14	4.31
Cambodia	0.31	0.50	0.34	0.09	0.03	0.24	1.51	10.56	9.05	0.78	8.28
China	0.39	0.12	0.21	1.18	0.04	10.53	12.46	8.36	–4.10	–0.33	–3.77
India	0.29	0.02	0.15	0.19	0.02	4.15	4.83	6.93	2.10	–0.10	2.20
Indonesia	0.32	0.15	0.28	0.87	0.03	6.39	8.03	12.54	4.50	0.93	3.58
Japan	0.48	0.20	0.61	4.09	0.09	47.73	53.21	8.77	–44.44	–3.62	–40.82
Korea DPRP	0.29	0.01	0.20	0.75	0.03	42.66	43.93	9.11	–34.82	–1.76	–33.06
Korea Republic	0.50	0.16	0.46	3.61	0.07	34.89	39.69	9.04	–30.65	–2.57	–28.07
Laos	0.27	0.76	0.58	1.28	0.03	0.34	3.26	23.81	20.55	4.64	15.91
Malaysia	0.59	0.30	0.44	4.04	0.04	30.07	35.48	17.10	–18.38	0.44	–18.82
Mongolia	0.38	15.05	0.26	0.01	0.01	11.05	26.77	189.89	163.12	25.17	137.95
Myanmar	0.41	0.05	0.39	0.73	0.03	0.98	2.58	11.99	9.40	1.06	8.35
Nepal	0.28	0.22	0.31	0.08	0.03	0.68	1.60	7.54	5.93	0.12	5.81
New Zealand	2.91	2.85	2.72	6.45	0.10	33.51	48.54	84.73	36.18	2.47	33.72
Pakistan	0.30	0.02	0.13	0.39	0.02	3.84	4.69	7.07	2.39	–0.29	2.67
Papua New Guinea	0.26	0.18	0.80	2.03	0.03	1.81	5.11	70.08	64.97	14.69	50.28
Philippines	0.27	0.17	0.32	2.13	0.03	5.63	8.55	7.99	–0.56	–0.45	–0.11
Sri Lanka	0.26	0.19	0.23	1.76	0.03	3.58	6.04	8.25	2.21	–0.36	2.57
Thailand	0.31	0.10	0.27	1.89	0.03	13.35	15.95	9.67	–6.27	–0.14	–6.13
Vietnam	0.27	0.06	0.24	1.21	0.05	2.31	4.12	9.14	5.02	0.20	4.82

TABLE 3.5C

Ecological Footprint per Capita in Global Hectares (EF-NPP Method) (Latin America, Caribbean, and North America)

World	Crop Land 0.52	Pasture 0.47	Forests 0.46	Fisheries 1.05	Built Space 0.05	Energy 19.36	Total FP 21.91	Biological Capacity 15.71	Ecological Balance (NPP) –6.20	Ecological Balance (GAEZ) –0.33	Difference –5.87
Latin America and Caribbean	0.49	1.95	0.57	0.64	0.05	13.20	16.90	22.22	5.31	1.67	3.64
Argentina	1.04	3.12	0.28	0.76	0.05	17.81	23.05	39.26	16.20	4.24	11.96
Bolivia	0.40	3.28	0.21	0.10	0.03	6.97	11.00	48.60	37.60	8.63	28.97
Brazil	0.66	2.92	1.02	0.59	0.05	8.88	14.11	29.16	15.05	4.67	10.38
Chile	0.52	1.46	1.85	1.25	0.08	14.28	19.44	39.84	20.40	1.41	18.98
Colombia	0.29	2.70	0.18	0.37	0.05	7.57	11.17	18.57	7.41	1.92	5.48
Costa Rica	0.38	1.42	1.05	0.28	0.08	11.11	14.32	18.85	4.54	0.58	3.96
Cuba	0.59	0.32	0.15	0.40	0.03	11.61	13.10	9.60	–3.50	–0.48	–3.02
Ecuador	0.35	1.59	0.73	1.00	0.04	9.86	13.56	15.92	2.36	1.03	1.33
El Salvador	0.34	0.92	0.55	0.20	0.03	5.53	7.57	7.43	–0.14	–0.59	0.45
Guatemala	0.31	0.87	0.67	0.04	0.05	5.69	7.64	10.33	2.69	0.25	2.44
Haiti	0.26	0.41	0.15	0.12	0.02	1.18	2.15	7.18	5.03	–0.23	5.26
Honduras	0.37	1.38	0.75	0.10	0.04	5.79	8.44	13.54	5.11	0.39	4.72
Jamaica	0.44	0.73	0.48	1.18	0.04	23.02	25.90	11.49	–14.41	–1.72	–12.69
Mexico	0.75	1.58	0.34	0.63	0.04	19.80	23.14	14.34	–8.81	–0.66	–8.15
Nicaragua	0.66	1.25	0.52	0.13	0.04	4.81	7.42	19.41	11.99	2.05	9.93
Panama	0.55	2.01	0.28	0.59	0.04	18.85	22.32	22.84	0.51	1.32	–0.81

Paraguay	0.47	3.34	1.19	0.55	0.04	4.27	9.86	34.37	24.51	5.44	19.07
Peru	0.43	1.20	0.19	1.48	0.05	3.71	7.06	30.11	23.05	5.16	17.89
Trinidad and Tobago	0.47	0.31	0.28	1.35	0.04	59.87	62.31	11.53	–50.78	–3.92	–46.86
Uruguay	0.58	8.12	1.00	1.29	0.04	11.60	22.64	44.42	21.77	4.58	17.19
Venezuela	0.43	2.06	0.14	1.04	0.05	25.08	28.80	19.75	–9.05	0.99	–10.04
North America	1.72	1.01	2.69	1.59	0.15	88.83	95.99	53.16	–42.83	–0.93	–41.90
Canada	1.90	1.06	2.74	1.31	0.12	75.91	83.03	85.95	2.92	3.12	–0.21
United States of America	1.53	0.96	2.65	1.86	0.18	101.76	108.95	20.37	–88.58	–4.99	–83.60

TABLE 3.5D

Ecological Footprint per Capita in Global Hectares (EF-NPP Method) (Western, Central, and Eastern Europe)

World	Crop Land 0.52	Pasture 0.47	Forests 0.46	Fisheries 1.05	Built Space 0.05	Energy 19.36	Total FP 21.91	Biological Capacity 15.71	Ecological Balance (NPP) –6.20	Ecological Balance (GAEZ) –0.33	Difference –5.87
Western Europe	0.98	0.53	1.78	2.87	0.11	54.45	60.70	16.84	–43.86	–2.73	–41.13
Austria	0.82	0.60	1.92	1.04	0.06	46.69	51.13	9.94	–41.19	–3.07	–38.12
Belgium and Luxembourg	0.85	0.44	0.00	2.02	0.13	65.43	68.87	7.19	–61.69	–4.64	–57.05
Denmark	1.26	0.45	3.28	1.69	0.11	55.04	61.84	16.28	–45.56	–2.64	–42.92
Finland	1.03	0.10	5.69	1.98	0.25	35.44	44.48	32.16	–12.33	2.12	–14.45
France	1.13	0.63	1.10	2.51	0.10	60.36	65.82	11.29	–54.54	–3.29	–51.25
Germany	0.73	0.22	0.87	1.22	0.14	49.04	52.21	8.44	–43.77	–2.75	–41.02
Greece	1.16	1.16	0.51	2.43	0.07	62.53	67.85	11.79	–56.06	–3.68	–52.38
Ireland	1.21	0.62	0.19	1.64	0.07	60.69	65.42	27.02	–38.40	–1.48	–36.92
Italy	0.86	0.49	0.69	1.75	0.03	37.68	41.51	8.05	–33.46	–2.65	–30.81
Netherlands	0.93	0.51	0.98	1.39	0.09	65.19	69.09	7.96	–61.13	–4.58	–56.54
Norway	0.76	0.16	2.40	5.91	0.09	83.81	93.13	48.89	–44.24	–2.90	–41.34
Portugal	0.93	0.66	1.02	10.03	0.09	36.47	49.20	16.33	–32.88	–3.58	–29.30
Spain	0.10	0.69	0.84	4.58	0.05	43.42	50.68	10.44	–40.24	–3.19	–37.05
Sweden	1.20	0.34	5.34	2.51	0.24	57.13	66.76	26.38	–40.38	–0.93	–39.45
United Kingdom	0.71	0.81	0.81	2.33	0.09	57.81	62.56	10.45	–52.11	–3.73	–48.38
Central and Eastern Europe	0.96	0.53	0.86	0.67	0.04	28.30	31.36	12.45	–18.91	–0.99	–17.92
Albania	0.65	0.44	0.13	0.20	0.04	8.43	9.90	8.29	–1.61	–0.55	–1.05
Belarus	0.89	0.75	0.45	0.42	0.03	32.62	35.17	12.23	–22.94	–0.79	–22.15

Bosnia Herzegovina	0.59	0.97	0.70	0.26	0.03	18.64	21.20	8.13	–13.07	–1.67	–11.40
Bulgaria	1.07	0.52	0.36	0.19	0.04	31.47	33.65	10.76	–22.88	–1.33	–21.56
Croatia	0.90	0.45	0.71	0.78	0.05	25.75	28.64	12.87	–15.77	–0.80	–14.97
Czech Republic	0.99	0.25	1.29	0.98	0.06	43.67	47.24	9.95	–37.30	–2.09	–35.21
Estonia	1.24	0.51	3.31	1.39	0.04	23.36	29.84	20.97	–8.88	–0.11	–8.77
Hungary	0.73	0.32	0.69	0.51	0.06	32.62	34.93	9.28	–25.65	–1.10	–24.55
Latuia	2.06	0.65	2.66	–0.12	0.03	18.83	24.11	18.75	–5.36	0.63	–6.00
Lithuania	1.40	0.36	0.82	2.33	0.04	32.56	37.51	12.41	–25.10	–1.22	–23.88
Macedonia	0.67	1.05	0.38	0.54	0.03	23.24	25.92	8.32	–17.60	–1.74	–15.86
Moldova Republic	0.64	0.24	0.09	0.15	0.03	8.37	9.52	7.73	–1.79	–0.21	–1.59
Poland	1.00	0.29	0.70	0.88	0.05	31.38	34.31	9.18	–25.13	–1.54	–23.59
Romania	0.78	0.67	0.42	0.19	0.04	26.84	28.94	8.24	–20.71	–1.65	–19.05
Russia	1.28	0.82	0.70	1.87	0.03	43.65	48.35	35.94	–12.42	0.69	–13.11
Slovakia	0.79	0.42	0.95	0.45	0.04	33.56	36.22	10.73	–25.48	–1.19	–24.29
Slovenia	0.70	0.42	0.88	0.48	0.05	36.05	38.57	10.88	–27.69	–1.34	–26.35
Ukraine	0.97	0.42	0.18	0.48	0.04	38.36	40.46	9.52	–30.94	–1.75	–29.19

of biocapacity in pasture and marine and inland fisheries (pafishr) and built space (bsshr) and the log of its total land area (lnarea) are employed as independent variables. EF-GAEZ shares were selected since the EF-GAEZ ecological balance is the baseline from which the ecological difference was calculated. For a given country *n*, the specified model is

$$diff_n = gdp_n + \ln area_n + enfpshr_n + pafshr_n + bsshr_n + e_n,$$

where *diff* is the difference in ecological balances as reported in Table 3.5, *e* is the error term, and all other variables are as defined above.

Table 3.6 reports the results. As expected, GDP, energy footprint share (enpfshr), and built space footprint share (bsshr) pull a nation's ecological balance down relative to EF-GAEZ, while land area (lnarea) and pasture/fisheries share (pafishr) push that balance higher. All variables were significant at the 0.05 level or better, with an adjusted R^2 of 0.5396. Energy footprint share has the strongest impact, while GDP has the weakest effect. Thus, our hypothesis that countries with higher energy consumption, energy footprint share, and built space share fare worse under EF-NPP relative to EF-GAEZ, while those with more land area, relatively

TABLE 3.6

Cross-Sectional Analysis of Ecological Balance Differences

Independent Variables	Coefficients
GDP	.0.71***
	(–3.00)
Lnarea	4.23***
	(3.02)
Enfpshr	–77.12***
	(–7.75)
Pafishr	24.70**
	(2.16)
Bsshr	–57.16**
	(–2.30)
Constant	–28.66
	(–1.52)
R^2	0.5396
Observations	134

Note: Numbers in parentheses are t-statistics.

, *p* = .05; *, *p* = .01.

larger shares of biocapacity in pasture and fisheries, and lower energy consumption fare better, is supported by our findings.

In some cases, differences between the approaches were significant enough to cause nations with positive ecological balances under EF-GAEZ to show negative balances under EF-NPP, and vice versa. For example, under our EF-GAEZ model as well as in Loh and Wackernagel (2004), the Russian Federation and Venezuela showed positive ecological balances. Under EF-NPP, these countries ran significant ecological deficits. The change in status was attributable, in part, to the rather large share energy consumption represented in these countries' EF-GAEZ footprints (56%). Conversely, under our EF-GAEZ model as well as in Loh and Wackernagel (2004), Mauritius and Yemen showed negative ecological balances. Under EF-NPP, these countries ran significant ecological surpluses. A key explanatory factor is EF-GAEZ pasture/fisheries footprint share, which was quite high in both at 63 and 46%, respectively. Such findings suggest that explanatory variables identified in our regression analysis may be important to any subsequent modeling completed with EF-NPP data.

Concluding Thoughts and Future Refinements

At a time when applications of ecological footprint analysis (EFA) are proliferating and international standards are under consideration, it is important to make changes to the basic methodology (EF-GAEZ) that responds to long-standing critiques that appear frequently in the literature. In this chapter, we made four. First, we expanded the purview of EFA to include the entire Earth in biocapacity accounts. Second, we made a formal accommodation for other species by deducting 13.4% of each biome from biocapacity. Third, we changed the way in which carbon sequestration rates and the resulting footprint are calculated. Finally, we changed the basis for equivalence factors from agricultural productivity potential to net primary productivity. Using the new approach (EF-NPP), we constructed ecological footprint accounts for 138 countries between 1961 and 2001 and compared our results with EF-GAEZ. A regression model was used to systematically analyze differences in ecological balances.

At a global level, key differences between EF-NPP and EF-GAEZ are apparent in the significantly larger biocapacity and ecological footprints shown by the former. In addition, EF-NPP shows ecological overshoot (negative ecological balance) for crop land, built space, marine and inland fisheries, and energy land, whereas EF-GAEZ reports overshoot only with respect to energy land, though no energy land biocapacity is explicitly included. Overall, EF-NPP shows a current ecological overshoot about

18% larger than EF-GAEZ. On a country-by-country basis, seventy-five countries had greater ecological balances under EF-NPP, while in sixty-three, the ecological balance dropped. Cross-sectional regression analysis supported the hypothesis that energy consumption, energy footprint share, and biocapacity shares for built space, pasture land, and marine and inland fisheries significantly affect the size and direction of ecological balance differences between the two approaches.

Our hope is that EF-NPP improves the accuracy and practical utility of EFA as a sustainability research tool. While EF-NPP represents a significant change, many more advances are warranted. One critical advance would be to establish sustainable yield benchmarks, as recommended frequently in ecological footprint literature. Mathematically, it is impossible for EF-GAEZ to show unsustainable use of forest land, crop land, marine and inland fisheries, or pasture land biomes on a global level due to the nature of the calculations and because EF-GAEZ fails to distinguish between sustainable and unsustainable yields (Ferng 2005).[8] Site-specific sustainable yield factors for fisheries and forests have been estimated by ecologists (Northcote and Hartman 2004; Ferng 2005). It may be possible to develop others by considering the connection between NPP removal and declines in biological diversity or the "species energy hypothesis."

According to this hypothesis, at levels above 50%, human appropriation of NPP has negative consequences in terms of biological diversity (Wright 1990). Haberl et al. (2004) recently validated the hypothesis in a study of agricultural lands in eastern Austria. With further research, it may be possible to use NPP retention standards embodied in a representative sample of sustainable agriculture, forestry, and fishery systems from various regions across the world as a basis for ecological sustained yield estimates applicable throughout entire crop land, forest land, pasture land, and marine biomes. From there we can convert the disparity between sustainable yields and actual yields reported by FAO into deductions from biocapacity.

For example, with respect to agriculture, organic agriculture methods retain considerably more NPP and, in a recent long-term study, have been shown to reduce crop yields by roughly 20% but enhance long-term soil fertility and biological diversity (Mäder et al. 2002). If FAO reported yields are presumed to be 20% above an ecological sustained yield, then the area needed to produce that excess yield can be converted into global hectares and then deducted from biocapacity. While reserved NPP is clearly not a sufficient condition for preserving biological diversity or ensuring sustainable land use in any particular place—a feat dependent upon many other local factors, such as the degree of fragmentation, presence of invasive species, and responsiveness of public officials—it is, nonetheless, a necessary condition and one that continues to be explored in a variety of ecosystems and settings. Moreover, because NPP provides a common foundation applicable to all EFA biomes, it appears to be well suited for incorporating ecologically meaningful sustainability standards into EFA. Incorporating sustained yield factors into EFA raises some challenges because data over time and across all biomes do not appear to be readily available. Nonetheless, sustained yield benchmarks are critical to the long-term effectiveness of EFA and ought to be thoroughly researched. As it stands now, EFA has a bias toward intensive (and potentially ecologically detrimental) production practices in the present over sustainability of ecosystems in the long run.

NPP can help fill other holes in EFA as well. Recent advancements in real-time satellite mapping of NPP (or its derivative measures, such as the normalized difference vegetation index (NDVI)) as well as human appropriation of NPP could also prove useful in estimating footprints for less productive lands and open oceans since these areas are not now linked to consumption data in the current EFA framework (Running et al. 2004). NPP can also be used to identify portions of aquatic and terrestrial ecosystems warranting removal from biocapacity due to specific threats or high species richness. Such mapping may prove to be a useful supplement to existing conservation needs assessments, such as global biodiversity hot spots or gap analyses (Bawa et al. 2002).

Additional refinements to EFA warranted by critiques and now being explored by the authors also include incorporating additional greenhouse gases, addressing the effects of environmental toxins, and modeling the footprint of water consumption. While carbon dioxide makes up the largest share of climate changing gases from anthropogenic sources, analyses that link methane (CH_4), nitrous oxide (N_2O), and fully fluorinated compounds (PFCs, HFCs, and SF_6) to appropriation of biocapacity would represent a significant step forward in making EFA more comprehensive and meaningful with respect to the effects of climate change. At first glance, it appears very difficult, if not impossible, to convert the impacts associated with uranium, lead, arsenic, mercury, and other toxics into an area-based measure such as footprint. Footprinting is, after all, a quantitative, not qualitative, indicator. Still, the relationship between concentrations of these toxins in a biome and its NPP may shed light on techniques to expand the scope of EFA to address these critical environmental concerns. Likewise, developing a defensible footprint for water consumption that captures aquifer depletion, loss of ecologically sustainable in-stream flows, and degradation of water quality would represent a significant improvement in accounting for vital ecosystem services performed by lakes, rivers, streams, and underground water reserves. Before international standards for EFA are promulgated, we believe the modifications discussed in this chapter as well as additional refinements should be fully explored.

Abbreviations

EF: Ecological footprint
EFA: Ecological footprint analysis
EF-GAEZ: Ecological footprint based on GAEZ suitability indices
EF-NPP: Ecological footprint approach that employs net primary productivity
EQF: Equivalence factor
ESY: Ecological sustained yield
FAO: United Nations Food and Agricultural Organization
GAEZ: Global agricultural ecological zone
GDP: Gross domestic product
GFN: Global Footprint Network
gha: Global hectare
Gt C: Gigatons of carbon
ha: Hectare
IPCC: Intergovernmental Panel on Climate Change
NPP: Net primary productivity
RP: Redefining Progress

Endnotes

1. Two clarifications are in order. First, the footprint does not provide a way to measure the environmental area impacted from pollution directly, so this should not be inferred from our definition. Second, the fresh water footprint does not account for human water consumption; it only accounts for our use of inland fisheries. The first version of footprint ever circulated in academic circles by William Rees and Mathis Wackernagel at the University of British Columbia in the 1990s did attempt to account for fresh water appropriation, but the approach was deemed inadequate and abandoned.
2. See www.myfootprint.org.
3. For example, Staples, an office supply company in North America, released its corporate sustainability assessment report, with an emphasis on the "environmental" footprint. However, the report presented raw descriptive data and no footprint calculations were performed.
4. See http://www.footprintnetwork.org/.
5. It is also worth noting that the FAO considers GAEZ data to be of uneven quality and reliability, and though various modes have been pursued for ground-truthing and verifying GAEZ suitability analyses, there is an acknowledged need for further validation of results and underlying databases (FAOSTAT 2005).

6. The EF-GAEZ model seems to show a net negative footprint for fisheries. This, given the model assumptions, appears to be an indication of an error in the model or in the data. It amounts to about 0.0056 gha per capita. This may not have been reported previously due to rounding or could, in fact, be a relatively minor flaw.
7. GDP is an indirect measure of the absolute magnitude of energy consumption in a nation while the share of energy in the overall footprint is a relative measure. We included GDP in lieu of a more direct measure of energy consumption due to multicolinearity concerns with energy share, and also because GDP is highly correlated with the overall size of the consumption footprint.
8. The proof is relatively straightforward. For crop land, a nation's footprint = [crop production (C) in tons per year/global crop yield factor (G) in tons per hectare per year] × the crop land equivalence factor (Q). Biocapacity = maximum crop land area (Am) × national crop yield factor (N) × the crop land equivalence factor (Q) (Wackernagel et al. 2005). We can factor Q out of both sides of the equation, leaving [C/G] on the footprint side and AmN on the biocapacity side. We can rewrite N as [Y/G] since N is simply the ratio of a nation's crop land yield (Y) to the global crop yield factor (G). For the world as a whole, however, Y = G, so this term is simply 1. This leaves C/G on the footprint side and Am on the biocapacity side. C/G cannot be greater than Am on a global basis because G is simply global production/global crop land area in production, or C/Ap, leaving Ap on the footprint side and Am on the biocapacity side, with Ap ≤ Am by definition.

References

Aall, C., and Norland, I.T. 2005. The use of the ecological footprint in local politics and administration: Results and implications from Norway. *Local Environment* 10:159–72.

Ajtay, G.L., Ketner, P., and Duvigneeaud, P. 1979. Terrestrial primary production and phytomass. In *The global carbon cycle*, ed. B. Bolin, E.T. Degens, S. Kempe, and P. Ketner, 129–81. Chicester, UK: Wiley.

Amthor, J.S. 1998. *Terrestrial ecosystem responses to global change: A research strategy*. ORNL Technical Memorandum 1998/27, Ecosystems Working Group of the Oak Ridge National Laboratory, Oak Ridge, TN.

Barrett, J. 2001. Component ecological footprint: Developing sustainable scenarios. *Impact Assessment and Project Appraisal* 19:107–18.

Barrett, J., Cherret, N., and Birch, R. 2004. *Exploring the application of the ecological footprint to sustainable consumption policy*. York, UK: Stockholm Environment Institute, University of York Heslington.

Bastianoni, S., Pulselli, F.M., and Tiezzi, E. 2004. The problem of assigning responsibility for greenhouse gas emissions. *Ecological Economics* 49:253–57.

Bawa, K., Rose, J., Ganeshaiah, K., Barve, N., Kiran, M., and Umashaanker, R. 2002. Assessing biodiversity from space: An example from western Ghana. *Conservation Ecology* 6:7.

Botkin, D.B., and Simpson, L.G. 1990. Biomass of the North American boreal forest. *Biogeochemistry* 9:161–74.

Chambers, N., Simmons, C., and Wackermenagel, M. 2000. *Sharing nature's interest: Using ecological footprints as an indicator of sustainability*. UK: Earthscan.

Cole, V., and Sinclair, A.J. 2002. Measuring the ecological footprint of a Himalayan tourist center. *Mountain Research and Development* 22:132–41.

Dias de Oliveira, M.E., Vaughan, B.E., and Rykiel, E.J. 2005. Ethanol as fuel: Energy, carbon dioxide balances, and the ecological footprint. *Bioscience* 55:593–602.

FAOSTAT. 2005. Data. Last updated February.

Ferguson, A.R.B. 1999. The logical foundations of ecological footprints. *Environment, Development, and Sustainability* 1:149–56.

Ferng, J.-J. 2005. Local sustainable yield and embodied resources in ecological footprint analysis—A case study on the required paddy in Taiwan. *Ecological Economics* 53:415–30.

Goessling, S., Hansson, C.B., Hoerstmeier, O., and Saggel, S. 2002. Ecological footprint analysis as a tool to assess tourism sustainability. *Ecological Economics* 43:199–211.

Gorham, E. 1995. The biogeochemistry of northern peatlands and its possible responses to global warming. In *Biotic feedbacks in the global climatic system: Will the warming feed the warming?* ed. G.M. Woodwell and F.T. McKenzie, 169–87.

Haberl, H., Schulz, N., Plutzar, C., Erb, K., Krausmann, F., Loibl, W., Moser, D., Sauberer, N., Weisz, H., Zechmeister, H., and Zulka, P. 2004. Human appropriation of net primary production and species diversity in agricultural landscapes. *Agriculture, Ecosystems, and Environment* 102:113–18.

Holden, E., and Hoyer, K.G. 2005. The ecological footprints of fuels. *Transportation Research: Part D* 10:395–403.

Hubacek, K., and Jiljum, S. 2003. Applying physical input-output analysis to estimate land appropriation (ecological footprints) of international trade activities. *Ecological Economics* 44:137–51.

Imhoff, M., Bounoua, L., Ricketts, T., Loucks, C., Harriss, R., and Lawrence, W. 2004. Global patterns in human consumption of net primary production. *Nature* 429:870–73.

Intergovernmental Panel on Climate Change (IPCC). 2004. Inter-annual and decadal variability of atmospheric CO_2 concentrations. In *The special report on land use, land-use change, and forestry*. http://www.grida.no/climate/ipcc/land_use/020.htm#table1-2.

Jackson, J., Kirby, M., Berger, W., Bjorndal, K., Botsford, L., Bourque, B., Bradbury, R., Cooke, R., Erlandson, J., Estes, J., Hughes, T., Kidwell, S., Lange, C., Lenihan, H., Pandolfi, J., Peterson, C., Steneck, R., Tenger, M., and Warner, R. 2001. Historical overfishing and the recent collapse of coastal ecosystems. *Science* 293:629–37.

Jorgenson, A.K. 2003. Consumption and environmental degradation: A cross-national analysis of the ecological footprint. *Social Problems* 50:374–94.

Kautsky, N., Berg, H., Folke, C., Larsson, J., and Troell, M. 1997. Ecological footprint for assessment of resource use and development limitations in shrimp and tilapia aquaculture. *Aquaculture Research* 28:753.

Lélé, S., and Norgaard, R.B. 2005. Practicing interdisciplinarity. *Bioscience* 55:967–75.

Levin, P., and Levin, D. 2002. The real biodiversity crisis. *American Scientist* 90:6.

Loh, J., and Wackernagel, M., eds. 2004. *Living planet report 2004*. Gland, Switzerland: World Wildlife Fund.

Mäder, P., Fliebach, A., Dubois, D., Gunst, L., Fried, P., and Niggli, U. 2002. Soil fertility and biodiversity in organic farming. *Science* 296:1694–97.

Mittermeier, R., Robles, P., Hoffman, M., Pilgrim, J., Brooks, T., Mittermeier, C.G., Lamoreux, J., da Fonseca, G., and Seligmann, P. 2005. *Hotspots revisited: Earth's biologically richest and most endangered terrestrial ecoregions*. Washington, DC: Conservation International.

Monfreda, C., Wackernagel, M., and Deumling, D. 2004. Establishing national natural capital accounts based on detailed ecological footprint and biocapacity assessments. *Land Use Policy* 21:231–36.

Northcote, T.R., and Hartman, G.F. 2004. *Fishes and forestry: Worldwide watershed interactions and management*. Oxford: Blackwell Publishing.

Opschoor, H. 2000. The ecological footprint: Measuring rod or metaphor? *Ecological Economics* 32:363–65.

Pauly, D., and Watson, R. 2001. Systematic distortions in world fisheries catch trends. *Nature* 414:534–36.

Post, W.M., Emanuel, W.R., Zinke, P.J., and Stangenberger, A.G. 1982. Soil carbon pools and world life zones. *Nature* 298:156–59.

Rees, W.E., and Wackernagel, M. 1994. Ecological footprints and appropriated carrying capacity: Measuring the natural capital requirements of the human economy. In *Investing in natural capital: The ecological economics approach to sustainability*, ed. A. Jansson et al. Washington, DC: Island Press.

Rodrigues, A.S.L., Andelman, S.J., Bakarr, M.I., Boitani, L., Brooks, T.M., Cowling, R.M., Fishpool, L.D.C., da Fonseca, G.A.B., Gaston, K.J., Hoffman, M., Long, J., Marquet, P.A., Pilgrim, J.D., Pressey, R.L., Schipper, J., Sechrest, W., Stuart, S.N., Underhill, L.G., Waller, R.W., Watts, M.E.J., and Yan, X. 2003. Global gap analysis: Towards a representative network of protected areas. *Advances in Applied Biodiversity Science* 5:73–74.

Rojstaczer, S., Sterling, S., and Moore, N. 2001. Human appropriation of photosynthesis products. *Science* 294:2549–52.

Rosa, E.A., York, R., and Dietz, T. 2004. Tracking the anthropogenic drivers of ecological impacts. *Ambio: A Journal of the Human Environment* 33:509–12.

Running, S., Nemani, R., Heinsch, F., Zhao, M., Reeves, M., and Hasimoto, H. 2004. A continuous satellite-derived measure of global terrestrial primary production. *Bioscience* 54:547–59.

Seyfang, G. 2003. Eco-warriors in the supermarket? *Evaluating the UK sustainable consumption strategy as a tool for ecological citizenship*. Norwich, UK: Centre for Social and Economic Research on the Global Environment, University of East Anglia.

Sonak, S. 2004. Ecological footprint of production: A tool to assess environmental impacts of tourism activity. *Journal of Tourism Studies* 15:2–12.

Sundquist, B. 2008. *Globalization: The convergence issue*. Edition 12. http://home.windstream.net/bsundquist1/gci.html (accessed May 26, 2009).

Torras, M. 2003. An ecological footprint approach to external debt relief. *World Development* 31:2161–71.

United Nations Environmental Program (UNEP). 2005. Indicators for assessing progress towards the 2010 target: Ecological footprint and related concepts. Paper presented at the meeting of the Subsidiary Body on Scientific, Technical and Technology Advice, Montreal, November 28–December 2.

University of British Columbia (UBC). 2008. Sea around us products. http://www.seaaroundus.org/ (accessed May 10, 2008).

van den Bergh, J.C.J.M., and Verbruggen, H. 1999. Spatial sustainability, trade and indicators: An evaluation of the ecological footprint. *Ecological Economics* 29:61–72.

Venetoulis, J. 2004. *City of Santa Monica ecological footprint 1990–2000*. Oakland, CA: Redefining Progress.

Vitousek, P.M., Ehrlich, P.R., Ehrlich, A.H., and Matson, P.A. 1986. Human appropriation of the products of photosynthesis. *Bioscience* 36:368–73.

Wackernagel, M., Monfreda, C., Moran, D., Wermer, P., Goldfinger, S., Deumling, D., and Murray, M. 2005. *National footprint and biocapacity accounts 2005: The underlying calculation method*. Oakland, CA: Global Footprint Network.

Wackernagel, M., Schulz, N.B., Deumling, D., Callejas Linares, A., Jenkins, M., Kapos, V., Monfreda, C., et al. 2002. Tracking the ecological overshoot of the human economy. *Proceedings of the National Academy of Sciences* 99:9266–71.

Wackernagel, M., and Silverstein, J. 2000. Big things first: Focusing on scale imperative with the ecological footprint. *Ecological Economics* 32:391–94.

Warren-Rhodes, K., Sadovy, Y., and Cesar, H. 2003. Marine ecosystem appropriation in the Indo-Pacific: A case study of the live reef fish food trade. *Ambio: A Journal of the Human Environment* 32:481–88.

Wolowicz, K. 2005. *The ecological fishprint of aquaculture: Can the blue revolution be sustainable?* Oakland, CA: Redefining Progress.

Wood, S., Sebastian, K., and Scherr, J. 2000. *Pilot analysis of global ecosystems: Agroecosystems*. Washington, DC: International Food Policy Research Institute and World Resources Institute.

Wright, D.H. 1990. Human impacts on the energy flow through natural ecosystems and implications for species endangerment. *Ambio: A Journal of the Human Environment* 19:189–94.

York, R., Rosa, E., and Dietz, T. 2003. Footprints on the Earth: The environmental consequences of modernity. *American Sociological Review* 68:279–300.

4

Learning by Experiencing: Systemics, Futures Thinking, and Scenarios

Richard Bawden and Michael Reichenbach

CONTENTS

Past performance is not necessarily a reliable indicator of future performance. This is true for all systems, whether economic, social, or ecological. Systems are dynamic. While we try to understand how changes in systems will affect sustainability, we cannot reliably predict outcomes, nor can we reliably select a desired future condition and control the systems to meet our future requirements—let alone our desired future condition.

Critical scenario learning for systemic development, or scenario learning, represents a framework for achieving sustainable development. This framework is especially helpful in situations where there is a high degree of uncertainty and complexity. If our current ways of being are not sustainable, how do we as individuals and as a society adopt new ways of being?

Before we can act differently, we must learn to see the world from different viewpoints, and from these new viewpoints take action. Scenario learning is a way of examining our own points of view and those of others.

Participants immerse themselves in the process of building scenarios about the future and learn from the process.

Both sustainable development and scenario learning are future and systems oriented. One of the principles of sustainable development shared by scenario learning is that those who are to benefit should be involved in the development of the solution. Another commonality shared between sustainable development and scenario learning is that they are processes. Participants learn from the ongoing evaluation of viewpoints in relation to a topic of interest. The conceptual underpinnings of the scenario learning process are based in systems and futures thinking (thinking about potential futures). This chapter presents the relationship of scenario learning to systems, futures thinking, and two approaches to scenario planning. One example of how the use of scenario learning resulted in actions toward restoring a forested ecosystem is presented. Scenario learning as a framework is well suited to application in sustainable development.

There is no doubt that our burgeoning human population is degrading and despoiling the environment, thereby depleting its finite resources at an ever-accelerating rate. It is both ironic and tragic that much of this circumstance represents the unintended consequences of the processes of industrial development. If we seek to sustain both our own well-being and that of the world about us, we need to reappraise our approach to the ideas of progress and development. We need to be much more reflexive about how we should be living our lives (Beck 1992, 260).

In essence, it is imperative that we adopt an ethic of sustainability as a vital context for everything we do. There is no sane alternative. As Simon Dresner (2002, 173) emphatically and poignantly argues in his book the *Principles of Sustainability*, "the alternative to the pursuit of sustainability is to continue along the present path of unsustainability, leading to disaster." This challenge calls for nothing less than a shift in our collective worldview from the narrowness and linearity of techno-optimism and political–economic short-termism to embrace a much more systemic and future-oriented perspective as the context for decisions about the strategic development of any of our human organizations.

The Call for Systemic and Futures Thinking

For all of the disagreement about the meanings and practical applications of the concept of sustainability—or what it means in both theory and practice to adopt an ethos of sustainability (Bawden 1997)—there is general agreement that thinking about *the potential future* is an essential orientation for thinking about sustainability. The essence of sustainability, after all, is persistence into the future, and our challenge is to figure out how to determine both what could and what should be allowed to persist (Thompson 2004). This fundamental question of sustainability asks how *systems* need to function in order to maintain their integrity through time. The advantage to thinking about sustainability in *systemic* terms and approaching it from a systemic perspective is that it helps us determine what could and should be allowed to persist.

A clear argument can also be mounted about the need for *futures thinking* as fundamental to the quest for sustainability, as Lester Milbrath (1989) argued two decades ago. His passionate but reasoned call was for us to commit ourselves to moving beyond what he persuasively argued was our unsustainable way of living, and to embrace the vision and work toward the creation of "a sustainable society" (Milbrath 1989). What was essential in this regard, he posited, was the formation of a "learning society" characterized by a form of "social learning" that would allow the reconnection "of society with its bio-community" in a manner that would be sustainable for both. And this, he further suggested, was a very different worldview or mental orientation from that which prevailed in the modern industrialized world.

Recognizing what he referred to as "systemic and futures thinking" as one of more than a dozen attributes of a learning society, he proposed the establishment of "special institutions where thinking systemically about the future is consciously practiced and methods are further refined for doing it well" (Milbrath 1989, 108). The implication was for the institutionalization of a paradigm—a revolutionary concept and a radical reorganization of the way people learn, such that the dominant worldviews held within societies could be challenged and transformed through a collective, conscious, and critical process of learning. While Milbrath failed to elaborate further on either the conceptual foundations or methodological implications of this profoundly unconventional way of thinking, he was clear in his inference that systemic and futures thinking was a single synthesis of what many would regard as two quite different and nontraditional perspectives—systems and potential futures.

Since Milbrath, few seemed to have taken up the particular paradigmatic or worldview challenge of either the synthesis of systemic thinking with futures thinking in the context of its application in the quest for sustainability, or its methodological refinement. Meanwhile, as Kunstler has recently opined, we are "sleepwalking into the future," seemingly unappreciative of the need for any urgency or even any necessity of questioning our assumptions about the kind of world we ought to live in or of the need to reflect critically on "the kind of world into which time and events are propelling us" (Kunstler 2005, 307).

There is an urgent need for change in the way we collectively learn about the world so as to facilitate fundamental changes to the way we currently live in it. A key complication is *what* each individual learns and *how* he or she comes to know it. The individual differences make the search for communal meaning, and thus consensual judgment, singularly difficult (Maturana and Varela 1988, 263). This difficulty is further compounded by the communication limitations imposed by language (Norton 2005, 607), and even more fundamentally by the apparent taboo of Western culture that "tells us it is forbidden to know about knowing" (Maturana and Varela 1988, 263). *Epistemology*, which is the study of the nature of knowledge, in particular its foundation, scope, and validity, is an important consideration in dealing with social–environmental sustainability. Learning about how we come to know and about the nature of knowledge itself is an important foundation for the sort of paradigmatic changes that sustainability dictates. As Bryan Norton readily concedes, the adoption of a social learning approach to environmental sustainability presents a host of philosophical as well as practical challenges to all concerned (Norton 2005, 607).

It is certainly true that some have recognized the significance of social learning to sustainable environmental management that, not unlike Milbrath's position, emphasizes the importance of a *systems orientation*, as well as of participation, integration, negotiation, and reflection (Keen et al. 2005, 3–21). In a complementary vein, the application of systems dynamics, systems analysis, research and design, and simulation modeling of both social and natural ecosystems (and sometimes even the integration of the two as socioecological systems) has achieved a particular prominence within the sustainability movement, and all are now regarded as key processes within the emerging sustainability sciences (Kates et al. 2001, 641–42). This has greatly facilitated the emergence and adoption of what has come to be known as adaptive ecosystem management (Holling and United Nations Environment Programme 1978, 377), which is also characterized by social learning and an embrace of contingency (Norton 2005, 607). Neither of these two important areas of intellectual and methodological development, however, has focused explicitly on a synthesis of *futures thinking* with *systemic thinking*, nor have they involved explorations of the cognitive implications of worldview challenges and change—or what, following Kitchener, might be referred to as the morality of intellectual development (Kitchener 1983, 222–32).

Initiatives in Systemic Development at the University of Western Sydney in Australia and at Michigan State University in the United States have revealed that part of the difficulty of connecting thinking about potential futures and systemic thinking lies with conceptual differences, as well as with intellectual and moral tensions between the two perspectives. This, of course, is all further exacerbated by the inherent ambiguities, contradictions, and contestable assertions that continue to characterize the concept of sustainability itself and its expression in practice as sustainable development (Bawden 2005, 151–64).

What is emerging from this work in progress is a promising social learning practice for exploring sustainability matters, which has explicit conceptual foundations in both systemic and futures thinking, and in reflexive and transformative experiential learning. While it does represent a response to Milbrath's call for methodological refinement from a synthesized perspective, it certainly has yet to provide an example of sustainable institutionalization (Bawden 2005, 151–64).

Systemic Thinking, Futures Thinking, and Social Learning

Before providing an account of a combined futures thinking and systemic learning methodology, it is important to elaborate on systemic thinking and futures thinking, respectively, and to introduce *social learning* as a reflexive and transformative experiential process. In the discussion that follows, each of these will be treated in a deliberately simplified manner in order to illustrate particular points of distinction. In practice, these distinctions are much fuzzier.

Systemic Thinking

To think systemically is to think in terms of bounded whole entities called systems, the interconnections and interrelationships within and among systems, the hierarchical organization of the systems, and emergent properties or unpredicted synergies associated with systems. From a first-order systems perspective, any natural or social organization, any community, or group of people can be perceived as a system that is composed of a set of interconnected subsystems. Every system is also itself perceived to be a part of a higher-order system, or a supra-system. Systems thinkers (systemists), therefore, think in and approach issues from the perspective of three spatial dimensions. They pay concurrent and integrated attention to the character and properties of each part of the system, the subsystems, and the supra-system. They are also interested in the interactions both within and among these dimension(s). Of particular importance to them are the unique properties that emerge through nonlinear interactions at every dimension. These interactions are unpredictable from a study of the isolated parts. It is these emergent properties that are the basis for the claim that systems are greater than the sum of their parts. It is this holistic belief or systemic assumption that fundamentally distinguishes systemic thinkers from nonsystemic or reductionist thinkers. Reductionists hold to the position that whole entities are not essentially different from their parts, that they can be known and understood from the independent study of the parts and of the causal and linear interactions and interrelationships that exist among the parts.

Systems that interact freely with the environmental supra-systems in which they are embedded are said to be open systems. The dynamics and nature of the flow of information within and among systems provides the negative feedback and positive *feed-forward* information that is essential to adaptation and coadaptation, and thus persistence into the future.

This basic open-systems notion of systems as whole, interconnected, equilibrium-seeking entities provides the conceptual foundation for the natural ecosystems that are the focus of many ecologists and environmental scientists. It is this ecosystems perspective that provides the logic of sustainability as a property of nature that is so often put at risk through human interventions. What is somewhat confusing in all of this, however, is the lack of clarity between ecosystems and the environment as the focus for sustainability. The place of human beings in natural ecosystems as well as in the environment also proves to be a point of considerable conjecture and uncertainty, for within ecology, human beings either are so frequently regarded as somehow unnatural components of natural ecosystems or are placed into such mythological and normative categories as the "Ecologically Noble Savage," the "Intruding Wastral," or the "Fallen Angel" (Berkes 1999, 209). Finally, of course, there are serious areas of disagreement about the systemic organization of nature that promote debates about the actual status of ecosystems in reality and whether they exist at all—about the true nature of nature as it were—and if they do exist, whether they naturally tend toward equilibrium (punctuated or otherwise), as is invariably claimed, or are inherently chaotic. All of this notwithstanding, there is no denying the usefulness of this type of first-order systemic thinking with its focus on systems as concrete entities in the real world, but it is not the whole systemic story.

The systems idea can be extended to embrace abstract or *second-order* notions of interconnectedness, for example, between thinking and valuing; rationality and emotion; objectivity and subjectivity; techno-science, ethics, and aesthetics; or even the knower and the known. Sets of ideas can be envisaged as coherent systems. The processes through which ideas are created, experiences are transformed into knowledge, or learners learn about learning itself can also be envisaged as coherent systems. Thus, it is possible and highly desirable for humans to experience, think, imagine, and act in such a manner that they essentially become systemic beings. This transformation requires considerable investment in intellectual and moral development that can be achieved only through prolonged and persistent challenge of how we learn and the nature of our knowledge (Bawden 2005, 151–64).

One of the apparent paradoxes of systemic thinking is that unless people appreciate second-order systemics, they are not able to work effectively in the domain of first-order systemics. To appreciate second-order systemics, experience with first-order systemics seems to be of considerable significance also. We find it hard to think of the world in systems terms or act in it in systemic ways until we learn to think about thinking in systems terms. These matters are all relevant to the strategic processes that will be explored later in this chapter. They also provide a useful framework for thinking about potential futures.

Futures Thinking

If systemic thinking is thinking in *three dimensions*, then futures thinking is thinking in a *temporal way*—placing a specific focus on what has yet to be or could possibly be within a context of what is of the present and what has been of the past. Simply put, it is thinking about potential futures.

In practice, there are two very different approaches with respect to how one thinks about the future, particularly as expressed in strategic approaches to planning for it. Each approach reflects a particular belief or set of assumptions about the *nature of the future* or at least about the best way to approach it. On the one hand, there is what might be termed the *getting-it-right* school of planning. Here the prevailing belief is that the best way to deal with the future is either to try to predict it, through a combination of extrapolation from the past and the present and of speculation about the future, or to design and actually try to create a future that is considered ideal, through deliberate manipulation of both internal and external circumstances. The future search method well exemplifies this planning to get-it-right school of strategy (Weisbord and Janoff 2000). It is a participative approach to the strategic development of an organization or a community, which progresses through a sequence of five simple tasks:

1. Reviewing the past
2. Exploring the present
3. Creating scenarios of the ideal future
4. Identifying common ground
5. Making action plans

Each task is typically the focus of a separate, participatory workshop or conference. These conferences are envisaged as social learning laboratories that are designed on participatory and action principles that, as future search proponents Weisborg and Janoff claim, lead to participants taking personal responsibility, fast implementation of action plans, and lasting relationships across key boundaries (Weisbord and Janoff 2000).

While considerable emphasis is placed on the importance of a perspective of *wholeness* to the process—whole person, whole community, whole world, and so on—and attention is drawn to the idea and usefulness of systems, there is usually no explicit reference made to either formal systems theories or systems practices during the conference process. In addition, there is no deliberate attempt made to nurture systemic thinking as a central aim of the process.

The primary focus of both the search conference and the future search method is on establishing a single ideal future of the organization or community of interest under review five, ten, or twenty years into the future. The methodology proceeds to the development of plans to create that ideal or desired future. Considerable attention is placed on activities that could be used to ameliorate and shape external forces to create the desired future. This might be considered an inside-out approach to the future. Essentially, the aim of future search is to learn about the characteristics of the ideal future, and about how that future might be achieved through manipulation of external variables.

On the face of it, there would seem to be considerable advantages in the application of the future search methodology to the practices of planning for sustainable development. It might be argued that the vast majority of initiatives in education for sustainability and of agendas for research and development are based on the design of ideal systems that can be managed in ways that are sustainable. Yet there is nothing explicit in the approach that provides a focus for the sort of knowledge or conceptual framework required to become systemic thinkers.

A second school of strategy takes a fundamentally different position with respect to the future: *planning to avoid getting it wrong* by taking an outside-in approach. Thus, in contrast to the getting-it-right school, the aim is the generation of a set of strategies to deal with different but plausible futures. While the process is also participatory and the emphasis is explicitly on social learning, there is typically a lack of methodological attention paid to the dimension of our knowledge or worldview that is proposed in learning from a potential future, which is at the heart of the quest for sustainability.

The scenario planning process is an example of a method within the avoid-getting-it-wrong school (Schwartz 1991). It normally consists of three stages of activities that, from a first-order systemic perspective, are concerned with identifying the issue of interest (within the system), generating a range of scenarios of the future states of the system's environment (at some agreed year in the future, usually around twenty to thirty years out) that are considered to be of strategic relevance to the issues identified for the system, and then using the scenarios as the contexts for the articulation of different adaptive and generative strategies.

More comprehensively, seven tasks are generally undertaken:

1. Identifying the focal issue of interest or concern to the organization or community
2. Clarifying the key factors that are likely to influence that issue
3. Identifying a comprehensive list of the driving forces for each of the following domains: natural, social, political, economic, cultural, and technological
4. Selecting and ranking the most critical of these driving forces with respect to their degrees of impact and uncertainty and expressing them in comparative states (high/low, fast/slow, adequate/inadequate, etc.)
5. Deriving the logic of different scenarios that reflect the different states of the critical drivers
6. Generating compulsive narratives of each of the different scenarios
7. Using the different narratives as contexts for testing the rigor of different strategies and for identifying early indicators of each scenario

Reference to systems thinking is often made within the scenario planning process, although the systems images that are usually employed are very basic and mechanistic. Characteristically, few, if any, attempts are made to explore the power and significance of systems thinking as an alternative way of viewing the world and the different states that it could plausibly assume in the future. The elaboration of the driving forces is typically conducted in a very linear manner.

There are a number of benefits of using a scenario planning approach within sustainable development practice. Its primary emphasis on the environment and on not getting the future wrong as the context for strategic actions contrasts sharply with the focus of the future search process on the designation of an ideal system of interest and the manipulation of the environment to achieve the ideal future.

Social and Experiential Learning

While scenario planning is all too often conducted in a mechanistic, instrumental, and indeed, nonsystemic manner, it is possible to conduct the process in an organic, systemic way that explicitly highlights the use of the technique as a process of reflexive, experiential, social, transformative learning. Although the conventional approach to scenario planning emphasizes learning about the characteristics of a number of different potential futures, it could be amended in a way that would exploit the possibility of learning experientially from those potential futures. A first step in that amendment would be the explicit adoption of experiential learning that was consistent with those aims.

The process of experiential learning is often portrayed as learning by doing, which greatly diminishes its significance to the process of sustainable development. A much more useful, relevant, and powerful way of looking at experiential learning is to appreciate it as learning by experiencing. Experience has an *inclusive integrity*, where there is no division between the act and the outcome; the subject and the object are "one unanalyzed totality" (Dewey 1938). In other words, experiential learning is essentially *learning by being*. Expanding on this theme, Kolb (1984, 256) argues that [experiential] learning "involves the integrated functioning of the total organism—thinking, feeling, perceiving and behaving." It is a vital process that involves iterative transactions between people and the environments to which they are constantly attempting to "holistically adapt." He formally defines it as the process by which "knowledge is created through the transformation of experience":

The basic premise of experiential learning is that human beings are involved in the continuous process of trying to make sense of their experiences, in order to take adaptive actions in it or coadaptive actions with it. During the process, experiences are transformed into knowledge, which is then used as the foundation for the actions that follow. In essence, the experiential learning process shifts back and forth between the world of concrete experience on the one hand and abstract concepts on the other. It also shifts back and forth between activities for finding out and those for taking action. When all of this is put together and integrated into a single systemic process, it can be presented as a cycle between the following four activities (or subsystems of cognitive processing): divergence/observing, assimilation/thinking, convergence/planning, and accommodation/taking action.

For Kolb, the process of experiential learning starts with the immersion of learners in some sort of problematic, disturbing, or curiosity-promoting concrete experience. This experience can be either real or imagined, from which as many observations and perceptions as possible are gathered, recorded, and shared. The goal of this activity is to create a comprehensive picture of the matter under review. When this picture is as rich as the learners would like it to be, they turn from the concrete to the abstract. They now reflect on what they have observed in an attempt to collectively understand what it is that they have experienced. This is an assimilation activity through which their observations are transformed into concepts; they make sense out of what they have observed. The third stage in the cycle moves from finding out to taking action—or, in this stage at least, about designing plans for taking action. This planning is when the thoughts about the matter at hand are further focused and interpreted. Finally, a decision is made to take action. To the extent that this action is evaluated in reference to the original problem, a new *cycle of learning* may be initiated.

While it is almost universally assumed that experiential learning is confined to the immediate present, as we each try to learn from what is currently happening to us, there is nothing to exclude the possibility of learning from experiences of a perceived future—a future that we can imagine into being with such richness that it becomes real. Furthermore, while considerable emphasis is usually placed on experiential learning as an individual activity, it can, with due care and attention, be adopted as a collective practice of social learning. These two possibilities are particularly useful in the process of critical scenario learning, which comprises an essential aspect of the systemic development approach to strategic development.

Critical Scenario Learning for Systemic Development

The basic proposition of critical scenario learning for systemic development is that people are much more likely to address the limitations of the way they live their lives if they experience the nature of multiple perceived futures. In essence, as a collective group of reflexive individuals, they can become a learning subsystem of the system of interest to them. What happens to the system in response to changes in its environment is going to personally affect those in the subsystem by virtue of them being embedded in both!

From what has already been said, it is clear that what is called scenario planning is much better understood as a process of learning—scenario planning—where "planning" is but part of a more systemically complete cognitive process that strongly interconnects finding out with taking action.

When conducted from within a perspective of systemic development, scenario planning has the capacity to be a systemic social process of reflexive, experiential, and transformational learning. Collectively, we can learn how to systemically transform our experiences of a range of different potential futures (that we have imagined into being) into knowledge that we can then use as the basis for informed consensual strategic plans and actions for dealing contingently with those potential futures. When conducted in a cognitively rigorous manner, we can collectively and systemically learn how to learn our way into the future. We can create powerful experiences of different potential futures, which we can use as strategic contexts to adapt to what we might have to face. When conducted with rigor, and in a critical manner, scenario planning can be transformed into critical scenario learning for systemic development. Scenario learning can help us transform the way we currently do things. Through attention to our own moral and intellectual development, we can learn to transform the way we view the world about us and to design subsequent actions that we should take to better live in it.

From the systemic development perspective then, a third critical dimension—the learning subsystem—must now be added to the focus on the systems of interest and on their potential future environments that are central to both the future search and scenario planning processes described earlier. Rather than the somewhat detached observers playing somewhat abstract planning roles for systems of interest in the future search and scenario planning processes, participants in systemic development strategic initiatives come to appreciate that they are vital components of those very systems. They must, therefore, do all that they can to experience what it is to be coherent subsystems embedded within those systems, with the responsibilities and capabilities for learning from the environments and for transforming their own experiences into knowledgeable adaptive or coadaptive changes that their studies of the future indicate for the system (as well as to themselves). They must approach the future with the belief that they can indeed imagine into being a range of plausible states of the environmental supra-system in which they, and the whole system of interest of which they are part, might well have to operate in the future. And they must develop the capabilities of not only generating those scenarios but also experiencing them in a manner that allows learning from them. Just as they represent the third dimension of the subsystem/system of interest/environmental supra-system, so too do they have their own three dimensions. They are capable of learning from the experiences they are imagining into being, learning how they are doing that learning, and learning about the nature of their own worldviews and the possible limitations that these are imposing on the other two forms or levels of learning. Borrowing from the logic and language of the analogous three-level model of cognitive processing developed by Kitchener, it is sensible to label these three levels as learning, meta-learning, and epistemic learning, respectively (Kitchener 1983, 222–32).

The Process of Critical Scenario Learning for Systemic Development

With these basic concepts in mind, it is now appropriate to outline the central process, and expand on some of the details of the practice of critical scenario learning for systemic development. Essentially the process is fashioned as three cycles of experiential learning—each comprising sequences of observation, reflection, interpretation, and decision making—conducted over a period (ideally) of several months and with each involving learning in three dimensions (learning, meta-learning, and epistemic learning). The output of the first cycle becomes the input for the second, and that of the second becomes the input for the third. There is, however, constant iteration, both within the cycles and among them, as the participants (the learning subsystem) reflect on the particular issues being addressed at any given time, the manner by which they are addressing those issues, and the nature of the worldviews that are prevailing within the learning subsystem and the impacts that those particular views might be having on the proceedings. This systemic development approach to scenario learning thus demands appreciation of the cognitive nature of experiential social learning and of the entire process as a reflexive and potentially transformative exercise. It also demands attention to the methodology itself as a meta-learning process and to the significance of limited knowledge with respect to constraints on learning or worldviews.

The overall aim of the process is not to try to predict what tomorrow will look like, or should look like, or what the participants would really prefer for it to look like. Rather, it is for them to learn how to imagine into being, and then experientially investigate a range of different but plausible future environments in which their organizational or communal system might have to live and work as the context for learning how to set strategies, both for reactive adaptation in the face of change and for proactive innovation to generate beneficial change.

It is important to emphasize that while the key learning strategy is experiential, there is also considerable reliance on knowledge or information (relevant to the issues being explored) that has been generated elsewhere by others.

Reflecting the three-level learning dimensions, there are three essential outcomes of the process: (1) the generation of scenarios themselves along with a number of robust strategies—both reactive or adaptive and proactive or generative—in response, (2) practice in the scenario learning methodology and additional understanding of it as an experiential and reflexive social learning process, and (3) appreciation of the nature of worldviews and their significance both to the scenario planning process itself and to learning in general.

Both experience and theory reinforce the claim that these meta- and epistemic learning dimensions of the scenario learning process have the most lasting and profound impacts on organizational, institutional, and community development, which is not to deny the importance of the scenarios themselves, nor the interpersonal fun (and more intangible organizational benefits) that comes with their generation and use.

The process of scenario learning is an iterative, reflexive experience. The knowledge that participants bring to the process and what they learn together influence their actions, thoughts, and future direction. It is essential that the process awaken within the participants the ability to think and work systemically. Learning to work systemically is predicated on epistemic learning, which is an awareness of one's own thought processes and of how the person's thinking is influenced by external factors. We also suggest that an awareness of systems is required for one to learn from the future and ultimately to work on a sustainable development project. The following will describe the three cycles. Cycle 1 will be described using an example from the North Shore Stewardship Association's experience with learning from the future. The process of learning from the future was applied to the sustainable development project of restoring what is known as the North Shore of Lake Superior's Lost Forest.

Cycle 1

Learning to identify focal issues of concern to a particular system involves (1) the identification of as wide a range of potential issues of significance to the system as possible from the past, present, and potential future—observation; (2) the synthesis of this rich picture of issues into a single theme—reflection; (3) the translation of this theme into one or two focal questions—interpretation—that reflect strategic intentions; and (4) testing the relevance of these questions with different categories of stakeholders, and refining them where indicated—decision. Over a period of eighteen months, participants in the Lost Forest project attended classes to learn about forests and forestry. They were motivated to gain knowledge about things and the processes required to restore the coniferous forest to the North Shore of Lake Superior. However, the program's organizers wanted participants to engage in the broader issue of sustaining the forest. The current forest is dominated by dying paper birch. Scattered coniferous trees, including balsam fir, white spruce, northern white cedar, red pine, and white pine, occur throughout the area. Forests that were dominated by these coniferous species were common in this area before 1900. Timber harvest, fires, and development contributed to the change in forest composition. Conifers are hard to establish due to deer browse. Sustaining this forest requires a view to the unknowable and unknown future; the ability to think about the systems involved, whether natural, social, political, economic, cultural, or technological; and the changing dimensions of each over time. The project, as described by John Meyers (2004, 1A, 5A) of the *Duluth News Tribune*, was to "start a small scale forest project that organizers hope will be the seed of big change along the shore, the first step in restoring the forest to what it looked like a century ago." The project organizers used a scenario learning process to start an ongoing discussion among participants. The objective of using scenario learning was to create a greater awareness of why each person thought and felt the way he or she did about the forests on the North Shore of Lake Superior. The setting was the unknown and unknowable future of 2050.

Participants discussed various perceptions of the North Shore forests using five domains: natural, social, political, economic, and technological. Participants were asked to record and discuss observations that might be made for each domain, starting with the present, 2005; looking to the past in 1960; and to the future of 2050. Observations were made as if the participant was living during the time of interest. For example, when examining 2050, participant statements would start with, "It is 2050 and I observe…." When investigating systems and developing plausible future scenarios, the participants found it helpful to look into the past. Lost Forest participants were asked to think about what the environment and life were like in 1960 on the North Shore of Lake Superior. Discussion was not limited to forests or forestry but included all aspects of life on the North Shore that each participant felt was relevant to the forest. Participants came to have a new view of life on the North Shore and how culture might affect restoration efforts. The participants came to realize that the North Shore is inhabited and used by longtime residents, longtime cabin owners, and tourists. Each group has its own culture and view of the forest. For example, tourists, and to some extent cabin owners, are likely to want paper birch replanted, whereas longtime residents are more apt to desire conifers. Looking into the past provides a means of learning about what drives change and provides a perspective on how much things can change over a forty-five-year period.

Participants were then asked to inhabit 2050 and record observations about the North Shore and the forest in each of the following six categories: natural, social, political, economic, cultural, and technological. For each observation, participants recorded assumptions and, where appropriate, causal agents or drivers for each observation. Participants often had differing views of the future. This was expected because each participant held differing worldviews. The goal was to learn from each other about viewpoints (system boundaries, facts and figures, other ways of perceiving, and so on).

The discussion catalyzed action on a broader scale than would have been achieved through classes on tree planting and forestry. Participants gained a greater awareness and an in-depth understanding of why the planting was important. Learning from the future connected these landowners to the issues on the North Shore and catalyzed their support and active participation in educational programs as leaders and teachers. One landowner took it upon himself to contact public agencies who own or control lands along the North Shore and encourage them to examine their forestry plans. Some volunteered to assist with forested-watershed-based education programs involving University of Minnesota Sea Grant and University of Minnesota Extension. One participant is now an instructor helping teach a class on intergenerational land transfer. These outcomes were not envisioned in the original work, but are outcomes of a dynamic learning process.

Cycle 2

Learning to generate scenarios for different plausible future states of the environment involves several steps. First, it requires the identification of a wide range of observations of potential environmental influences on a system, such as forests of the North Shore. Second, it requires the synthesis of these observations into forty or fifty key critical influences. These observations are then categorized and further reduced in number as a function of their assumed criticality. Third, it requires the integration of these critical influences into a number of different scenario logic sets of possible futures that differ according to the different critical influences and their interactions. Fourth, it requires the expansion of these logic sets into narratives that both describe the respective future worlds each represents and explain their (imagined) historical evolution in a manner that reflects the rigor and internal consistency of the scenarios that have been generated.

One scenario might describe an extreme future world (in 2030, for instance) in which climate change (natural domain) has accelerated dramatically as violent weather extremes become the norm, with coastal flooding accompanied by inland droughts. This has such a significant depressing effect on agriculture that food shortages, and thus malnutrition and starvation, are commonplace. Immunity to disease among populations falls accordingly and newly emerged, viral pathogens wreak havoc. Food riots increase in frequency, and a general breakdown of law and order is a constant threat (social domain). Health services are overwhelmed. The ineptitude of central governments increases as their ability to deliver crucial services is stretched way beyond the resources they need to accomplish the tasks (political domain). Economic conditions (economic domain) are harsh because the costs of natural resources, especially oil, have escalated in the face of both supply shortage and greatly increased demand from emergent industrializing nations. The associated growing poverty further exacerbates the effects of the food crisis while greatly amplifying social unrest. Fear, anger, and uncertainty due to the extreme conditions fuel individual competitiveness and aggression (cultural domain), and a culture of intolerance and mistrust prevail. Individualism is not only rampant but also reflected in the ever-increasing emphasis on fundamentalist perspectives within organized religions. While developments in nanotechnology, information, and biotechnologies (technological domain) have continued to advance since the early part of the twenty-first century, they are of little use in dealing with our human-caused changes in the weather and, indeed, with transnational terrorists employing bioagents and weapons of mass destruction in their tactics, technology has come back to bite the industrial modernists.

A contrasting scenario would see a much more benign world in which climate change (natural) has turned out to be far less severe and much more predictable in its trajectory than was imagined back in 2006. This has allowed societies (social) to make significant and timely adjustments to the way they operate, which is seeing an orderly transition back to the communal ways of living that characterized preindustrial life, but with many innovative conveniences making this much less harsh than the earlier era. There is a marked shift in patterns of governance (political), as localism replaces centralism in so many aspects of life. Localism has also extended to employment patterns and economies (economic) in many instances, as nonindustrial ways of food production have reappeared and are calling on the reestablishment of the myriad local businesses and services that had previously been taken over by corporations. Cultures (cultural) have become characteristically communal, and there is a strong resurgence in both community self-help and local philanthropy. This communalism has both triggered and benefited immensely from a marked shift away from fundamentalist religious positions to more community-oriented, softer religions with an emphasis on inclusive well-being, which co-exist harmoniously with a similarly soft secular humanism. Such inclusivity embraces respect for other, nonhuman life forms, and an ethos of sustainability has come to prevail in many places. There is an important fusion of spirituality with science, which has led to a greatly enhanced respect for technologies with a human scale. It is developments within such a human-faced technological culture (technological) that have seen vital pioneering initiatives in a whole host of critical domains—from alternative energy to medicine.

Cycle 3

Using the scenarios to learn and develop response strategies involves becoming familiar with the details of the scenarios, testing the utility and inadequacies of current plans based on the knowledge gained from the different scenarios, designing new plans in response to the different scenarios, and exposing the newly generated plans, along with their scenario contexts, to critical review. With a range of very different future scenarios to consider, the learning subsystems must now plan and select appropriate strategies that will enable their systems to adapt and respond if they are to persist into the future. Based on the review of the scenarios, they accept the responsibility of changing their current activities and designing strategic actions that might be capable of influencing the future.

A Final Word

It is important to emphasize that each cycle of learning is conducted by groups of people who are consciously assuming a second-order systemic perspective on themselves as a critical learning subsystem of an identified system of interest. With practice, participant groups come to adopt holistic perspectives on everything that they do. They start looking for interconnectivities, for interrelationships, for wholeness and integrity, and perhaps most significantly, for emergent surprise. The transformation from symptomatic thinking to systemic thinking is the most vital aspect of this entire endeavor—one that by far represents the best way for us to move toward a well-being of all generations that is both socially and environmentally sustainable.

References

Bawden, R. 1997. Learning to persist: A systemic view of development. Paper presented at Systems for Sustainability: People, Organizations and Environments: Proceedings of the Annual Conference of the UK Society for Systems Sciences, Milton Keynes, UK.

Bawden, R. 2005. Systemic development at Hawkesbury: Some personal lessons from experience. *Systems Research and Behavioral Science* 22:151–64.

Beck, U. 1992. *Risk society: Towards a new modernity*. London: Sage Publications.

Berkes, F. 1999. *Sacred ecology: Traditional ecological knowledge and resource management*. Philadelphia: Taylor & Francis.

Dewey, J. 1938. *Experience and education*. New York: Kappa Delta Pi. Republished by MacMillan Publishing, 1963.

Dresner, S. 2002. *The principles of sustainability*. London: Earthscan Publications.

Holling, C.S., and United Nations Environment Programme. 1978. *Adaptive environmental assessment and management*. Vol. 3, International Series on Applied Systems Analysis. Laxenburg, Austria: International Institute for Applied Systems Analysis.

Kates, R.W., W.C. Clark, R. Corell, J.M. Hall, C.C. Jaeger, I. Lowe, J.J. McCarthy, et al. 2001. Sustainability science: Environment and development. *Science Paleoclimate* 292:641–42.

Keen, M., V.A. Brown, and R. Dyball. 2005. Social learning: A new approach to environmental management. In *Social learning in environmental management: Towards a sustainable future*, ed. M. Keen, V.A. Brown, and R. Dyball, 3–21. London: Earthscan.

Kitchener, K.S. 1983. Cognition, metacognition, and epistemic cognition. A three-level model of cognitive processing. *Human Development* 26:222–32.

Kolb, D.A. 1984. *Experiential learning: Experience as the source of learning and development*. Upper Saddle River, NJ: Prentice-Hall.

Kunstler, J.H. 2005. *The long emergency: Surviving the converging catastrophes of the twenty-first century*. 1st ed. New York: Atlantic Monthly Press.

Maturana, H.R., and F.J. Varela. 1988. *The tree of knowledge: The biological roots of human understanding*. 1st ed. Boston: New Science Library.

Meyers, J. 2004. Fresh start for the forest. *Duluth News Tribune*, November 30.
Milbrath, L.W. 1989. *Envisioning a sustainable society: Learning our way out*. New York: State University of New York Press.
Norton, B.G. 2005. *Sustainability: A philosophy of adaptive ecosystem management*. Chicago: University of Chicago Press.
Schwartz, P. 1991. *The art of the long view: Planning for the future in an uncertain world*. New York: Currency Doubleday.
Thompson, P.B. 2004. Sustainable agriculture: Philosophical framework. In *Encyclopedia of plant and crop science*, ed. R.M. Goodman. New York: Marcel Decker, pp. 1198–2000.
Weisbord, M.R., and S. Janoff. 2000. *Future search: An action guide to finding common ground in organizations and communities*. 2nd ed. San Francisco: Berrett-Koehler.

5

Using Lean Thinking Approaches to Speed Sustainable Development

James R. Sinton

CONTENTS

Introduction

"Lean thinking" or *"thinking Lean"* is the passionate effort to eliminate waste, which, in turn, is any activity that uses resources without adding

social value. Here, two questions are important: (1) What has the concept of Lean Thinking got to do with sustainable development, and (2) why is the relationship worth exploring? The need for a radical switch to sustainable development is growing daily. As we remove more and more of the options for sustaining human life on Earth, it is becoming clear that only widespread and dramatic change toward sustainability will save our descendents from a poor quality of life. This being the case, every approach or method for rapidly and reliably transforming the way we live and work is worth investigating.

When we view human activity across the globe, we find one thing underlying nearly all of it—the pursuit of economic interests. It has become evident over the past half century that society will change faster through altering business practices and incentives than through any other means. It is also clear that business and industry are one of the primary engines of the transformation that will take our world into a sustainable future.

Lean Thinking (hereinafter simply referred to as "Lean") has emerged during the last ten years as the approach most likely to be around in business and industry at the end of the century. That makes it worth investigating.

There is no doubt that we are in the early stages of an unprecedented, worldwide paradigm shift toward sustainability (Edwards 2005). This is heartening news and certainly gives us hope that we will emerge from this transformation happier and wiser. It is, however, too early to celebrate.

Although humanity may have turned a page in the unfolding of its own development and its relationship to the universe, turning that page has not established the required changes at a speed sufficient to avert serious malfunctions in the way we live. The complex flow of materials and information that we call civilization is being stretched across an ever-widening abyss between the way nature does things and the way we do them.

As the paradigm shift spreads across an awakening humanity, the need for rapid and enduring change is becoming apparent, even within corporations, which are organizations that incorporate groups of people and systems. Here, the paradigm shift is causing much of the business world to confront an increasing pace of change, competition, and complexity. Driven by the need for rapid change, corporations have become a hotbed of transformative methodologies.

To the extent that a corporation is able to act as a single learning entity and adapt to a changing commercial environment, it is successful. Those that have embedded Lean into their operations and management are proving to be some of the most adaptable over the long term. Among these is Toyota, which is becoming the premier corporation in the auto industry (Liker 2004).

Large corporations have many employees who form distinctive communities within the corporate structure. Without consciously realizing it, management personnel have learned how to effectively engage all these communities in the Lean transformation as they pursue their quest for profits.

Lean has been compared to all the other major transformative approaches by different consultants and found to be the most cost-effective, enduring, and powerful. The author completed a survey of 102 midsize to large U.S. companies in the year 2002 in order to ascertain how many were engaged in Lean. Over 50% were either already engaged or planned to be within a year because they viewed Lean as the best way forward.

This industry-wide acceptance, plus the completeness of Lean, makes it an ideal choice as a framework to use for transforming organizations and communities into more sustainable entities. As awareness of the necessity for social transformation spreads, Lean can be modified and expanded so that businesses and communities seeking social-environmental sustainability have a mutually familiar framework in which to engage.

In this chapter, two major examples of Lean and sustainability will be examined: the U.S. Environmental Protection Agency "Lean and the Environment" Initiative and the Interface, Inc., Sustainability Initiative.

Understanding Lean Thinking

To reiterate, "Lean Thinking" or "Thinking Lean" is the passionate pursuit and elimination of waste, which, in turn, is any activity that uses resources without adding social value. Imagine our world with all the waste removed; we might work a four-hour week in order to deliver the same value that currently requires a forty-hour week. Clearly, that is not how it will end up, but the possibilities are endless.

The point is that Lean will remove the wasted resources as well as the wasted activities. When this is applied in the form of radical resource productivity, it is expected to yield between ten and twenty times more from the raw materials used in manufacturing. This idea has the undivided attention of many industries because the potential profits from such yield increases are staggering. Although Lean could thus help reduce the unnecessary use of nonrenewable resources, it can offer a lot more than this.

Lean is currently transforming more organizations in more countries than any previous approach to business improvement. Why? Lean represents the distilled wisdom and application of some of the greatest minds in the business (Lean draws on the work of Henry Ford, W. Edwards Deming, Taiichi Ohno—ex-president of Toyota, and many others). But more than that, Lean represents the accumulated refinement of methods tried in a wide variety of environments over many decades by thousands of organizations.

Lean started to crystallize with the work of Taiichi Ohno at Toyota, beginning in the 1950s, but was first packaged as a complete approach in 1996 by James Womack and Daniel Jones in "Lean Thinking" (Womack and Jones 1996). Because this led to an explosion of interest in Lean from every sector, it is now encountered in many service industries, including banks, hospitals, and hotel groups, as well as in many aerospace and automobile manufacturers, local and state governments, and even in the U.S. army.

Lean comprises a set of five principles, which are applied with four rules and at least twenty tools. These Lean elements are frequently combined with Six Sigma (a statistical-based methodology for removing unwanted variances from any process) and customized to suit the organization using them.

Using the Approaches of Lean Thinking for Sustainability

Lean can be used to facilitate sustainability in different ways. First, it can be used purely as a means to achieve greater efficiency and effectiveness. Second, it can be kept as pure Lean thinking, but accompanied by a broader definition of value that includes elements of sustainability. Third, we can borrow many of the principles, rules, and tools of Lean to use in projects aimed primarily at improving sustainability. These ways of using Lean are termed level 1, level 2, and level 3, respectively.

Level 1—Using Lean to Remove Waste

As we have seen, the removal of waste includes a more effective utilization of resources. Converting raw materials into commodities, like electric power, frequently entails a long chain of steps such that exceedingly little of the original resource is available to the end user, most having been lost along the way. In the case of electric lighting, we usually enjoy less than 10% of the original power embodied in the raw material. The monetary savings by the person who turns on the light at the end of the chain are thus magnified as we move upstream toward the raw material. In the case of electric lighting, a savings made by the end user results in a substantial reduction in the extraction of coal or oil and, consequently, a substantial reduction in emissions of carbon dioxide.

There are many other examples of Lean waste removal that have resulted in improved sustainability. For example, reduced industrial accidents and spills have reduced pollution; reduced space utilization cuts down on energy utilization; reduced defects lead to fewer discarded products; the service and flow model in which manufacturers lease or rent products means they recover and recycle them many times before disposing of them; in addition, increased product durability and life span have significantly reduced both resource utilization and the area needed for garbage dumps.

Increasingly effective resource productivity, one of the spin-offs of Lean, has already cut huge amounts of resource extraction in many industries, and it has hardly begun. Because it frees up large amounts of capital, radical resource productivity is being targeted before other types of sustainability. Fortunately, the resulting increase in available capital will help fund the broadening of the transformation to cover other types of sustainability.

Level 2—Lean Is Driven by the Definition of Value

The perception of value is the bedrock of Lean. If Lean is the removal of waste and waste is the absence of value, then Lean is driven entirely by how we define value. Most authors and consultants describe value as that which the customer is happy to pay for. In doing this, they are following the classic business model and missing a huge opportunity. If value is defined as that which all the stakeholders desire, rather than focusing solely on what customers desire, it widens the scope of Lean enormously.

After customers, the next most powerful stakeholders are the owners or shareholders. Beginning around 1990, the growing concern about issues of social–environmental sustainability has made these topics significant elements in the attraction of shareholders and equity partners, which forces the implementers of Lean to consider both social and environmental issues.

Other stakeholders, whose desires must be taken into account when defining value, are the employees, suppliers, and community in which the business operates. Finally, the ecological integrity of the environment is the most important consideration of all because it sustains both the community and the business. Moreover, when the definition of value includes both the community and environment, a business that was solely for profit is converted into an initiative for sustainability without having to tinker in any way with Lean.

This shift toward social–environmental sustainability happens unknowingly in many cases, where the team implementing the concepts of Lean includes one or more individuals who are aware of critical environmental or social issues. They slip some of this concern into the transformation process (known as Kaizen, which is Japanese for change and good) and so start moving toward a sustainable future.

It is, however, much more powerful when Lean is driven from the start by the definition of sustainable value. Every Lean rule and tool uses the definition of value as its baseline, thereby redefining the target toward which both sustainability in its widest sense, and profit in its narrowest sense, are moving. This is true because, in most cases, sustainability reduces expenses and improves sales.

Using a blend of values, such as social vs. environmental, can be complicated by the fact that the impact of different value elements on different processes in a value stream may vary. It is thus important to understand how a blending of values works.

The first challenge is to identify the stakeholders. Next, a common language and system of metrics must be found through which a blended value can be obtained. Even then, it will be necessary to assign different weightings to each element of the blended value so the weightings can be taken into account when the specific process under consideration is being mapped and its characteristics determined.

Let us take an example of a blended value that illustrates how powerful this approach is. During 1975, long before Lean was known, 3M Corporation initiated a program to reduce costs and prevent pollution. Although this program predates Lean, it uses most of the Lean approaches and is known as 3P (Pollution Prevention Pays). It targeted both the reduction of costs and the elimination of the causes of pollution, rather than simply its control. In Lean terms, they set out to remove waste by using a design strategy like the Lean "design for manufacture and assembly" tool. Effectively they included a high level of environmental value in their blended-value definition. Thirty years later, 3M celebrated their successes; "… the program had prevented more than 2.5 billion pounds of pollutants and saved over $1 billion based on aggregated data from the first year of each 3P project" (3M Corporation 2007). Specific examples of 3M's successes can be found on the 3M website (www.3m.com) under sustainability and Pollution Prevention Pays (3M Corporation 2008).

Level 3—Switching from Commercial-Lean Thinking to Sustainable-Lean Thinking

Lean thinking is based on five principles, four rules, and numerous tools. It is implemented using a transformation methodology known as Kaizen, which is based on continuous incremental improvement. Many of these Lean elements can be easily adapted for use in transformations to sustainability. By using previously tried and tested methods and approaches, we inherit many of their benefits. Examples of this translation process to create a new proposed framework for sustainable development are discussed in the next section.

Translation of Lean into Sustainable Lean

Lean Concepts

The concept of *waste* is a core driver of Lean. Waste refers to any activity or use of resources that does not add value as perceived by the end customer. The concept: *the removal of nonsustainable elements* is a core driver of sustainable Lean. This refers to any activity or use of resources that neither maintains nor enhances the sustainable value of the system.

The concept of *value* in Lean is defined precisely by what a customer is happy to pay for, specifically excluding any activity or resource the customer considers nonessential. Sustainable *value* refers to the balanced combination of those qualities that enhance the sustainability of any product or process in terms of ecology, economics, the surrounding community, or society as a whole. Such balance can be achieved through a process of blending the intrinsic values held dear by each of the stakeholder groups.

The concept of *value stream* refers to the sequence of accumulated values that are added to a design or concept and accumulate as the product or service becomes increasingly manifested, eventually providing the intrinsic value sought by the end customer. *Sustainable value stream* refers to the sequence of accumulated sustainable values that are added during the process of increasingly manifested sustainable value within a system.

The concept of *flow* refers to the continuous movement of a product or service created by balancing the work along the value stream in such a way that a single piece or element is produced at a time and flows along the value chain. This requires use of the Lean tool line balancing, which can be defined as the process of assigning exactly the right amount of resources at each step in the value stream so that flow is maintained. *Sustainable flow* emulates the tendency of natural systems to move materials and processes (including information in human systems) through their structures in a balanced way, which maintains their required levels, thereby preventing a buildup of unnecessary materials while ensuring that appropriate levels are always available for the next step in the cycle.

The concept of *pull* means that a product or service is produced only when a customer demands it, which ensures that customers get exactly what they want when they want it. *Sustainable pull* emulates the tendency of natural systems to move materials through their structures only as they are required by the next step in the cycle.

The concept of *perfection* is an ideal based on the understanding that it is a constantly moving management target, and so, while never achievable, is always worth pursuing. *Sustainable perfection* recognizes that change is a constant process configured in a curvilinear spiral, which, while cyclical in motion, can only approximate its beginning, never touch it. As such, change is the creative process that keeps the world ever novel, interesting, and evolving. Understanding this dynamic helps us to figure out how we can harness the force of evolution when designing artificial systems that emulate nature.

The Lean practice *Kaizen*, the engine of Lean, comes from the Japanese for "change" and "good." Kaizen actually mimics the process of meditation, where an individual shuts down the sensory inputs and allows the intelligence deep within the mind to work on the limiting factors responsible for egotistical behavior. *Sustainable Kaizen* is a process of continuous, incremental improvements that reengineers existing systems to align them more closely with the biophysical principles that govern nature and thereby eliminate waste. Sustainable Kaizen is thus the engine of sustainable Lean.

Lean Rules

Lean rule 1—standardize all processes—instructs those who implement Lean to examine, and where possible, standardize and continually refine every process to reflect the ongoing removal of waste. Sustainable standardization necessitates the creation of performance and operating standards for every repeated activity, both in the implementation process and in the system being transformed.

Lean rule 2—minimize unnecessary movements of people—urges the removal of every human movement that does not add value. Sustainably minimizing unnecessary movements by people means to reduce unnecessary human movement while also taking into account the values of all stakeholders. For example, designing or redesigning communities to minimize the necessity of motorized transportation can substantially reduce both the carbon footprint and the cost of doing business.

Lean rule 3—minimize the movement of materials—requires the removal of every movement of a material that does not add intrinsic value to the final product or service. Sustainably minimizing the movement of materials works as long as the measure of wasted movement concurs with the basic values of all the stakeholders.

Lean rule 4—educate everyone in the appropriate way at the appropriate time—seems obvious, yet is seldom applied, despite the demonstrable fact that time and money are continually wasted by providing too much information at inappropriate times. Sustainably educating everyone, appropriately and at the appropriate time, means that the concept of teaching through lectures is largely replaced by learning through engagement and experience. Here, a Chinese proverb is helpful: I hear and I forget; I see and I remember; I do and I understand.

Lean Tools

Lean tool 1—mapping the value stream—is the core tool for implementing Lean. A cross-functional team follows the value stream from the end customer toward its sources. They map each step in the process, taking note of the value added or not added, as well as the time that elapses with and between operations. This map is used to identify areas where the removal of waste can be improved. Maps are then created, which depict the current status and future possibilities for each value stream. Mapping the sustainable value stream is based on sustainability, which can be maintained only when the economic, ecological, and social aspects of a system remain in balance with one another. Mapping the sustainable value stream follows the development of these aspects, commencing at the product or most manifest end of each value stream and moving toward its multiple sources. These maps are then used to identify nonsustainable elements.

Lean tool 2—design for manufacture and assembly—brings together representatives of the entire value stream to ensure that the design process incorporates every value-added step, minus the identified waste. Designing for overall sustainability requires representatives from all parts of the sustainable value stream to work together during the design phase of any project, system, service, or product to ensure that all the elements of sustainability are included, while all the unnecessary steps and resources are excluded.

Lean tool 3—Total Productive Maintenance—uses all relevant organizational resources in the most effective and efficient manner to achieve the highest possible performance from machines and system dynamics over their entire life cycle, thereby ensuring the ability of the organization to deliver what the customers want when they want it. Total sustainable maintenance uses people from all relevant sections of the sustainable value stream working together to plan and execute a maintenance program that includes the proper functioning of individuals, teams, and communities, and thus ensures the most sustainable operation of the entire system.

Lean tool 4—visual workplace—renders critical communication effective among all relevant parties. Lean organizations display all critical information, such as standards, performance, and targets, on appropriate whiteboards, bulletin boards, or electronic devises. This practice replaces many of the cumbersome and wasteful computer reports used in non-Lean companies. Sustainable visibility ensures the effective communication of critical information among all stakeholders, which translates into the display of clear, understandable metrics and messages in appropriate places, where they are easily accessible to those who wish to use them.

Lean tool 5—Kanban—is a Japanese term based on the words *kan* for "card" and *ban* for "signal." Kanban is a replenishment process, where each production lot has an assigned card. Whenever a lot is consumed, the card is sent to the supplying work center as a signal for reorder. Sustainable Kanban is any system used to signal the need for more elements or value-adding activity in such a way that only the minimum essential levels of that element or activity are maintained without jeopardizing the continuous flow of the value stream.

Lean tool 6—error proofing—is also known as mistake-proofing or self-check systems. It assumes that humans will always make mistakes and so aims to prevent them rather than fix them. Common examples of error proofing are the childproof caps on medicine bottles that keep children from taking potentially harmful medication and the brake-shift interlock device on your automatic transmission vehicle, which prevents you from starting the engine unless the brake is depressed and the shift selector is in park or neutral. In Toyota, error proofing is used so successfully to ensure quality that quality assurance inspections have been eliminated, and vehicles are shipped abroad without ever testing the engines.

Sustainable error proofing mimics nature, which continuously evolves perfect systems for preventing errors. An example of natural mistake proofing can be found in the ways damaged DNA repairs itself. The base pair sequence coding is automatically repaired through referral to the mirror image DNA strand. Sustainable error proofing seeks to incorporate natural systems that have already evolved the self-check mechanisms that ensure consistency, rather than trying to control the outcome through human effort or high-tech solutions. An example of this is the use of companion planting for higher yield and pest control.

Lean tool 7—just-in-time—is a strategy for exposing waste and ensuring that continuous improvement occurs. It relies on total employee involvement and focuses on delivering what customers want, when they want it, in the quantity they want, and to the exact location they want. Sustainable just-in-time harnesses the power of nature to ensure that the material or action needed is provided at exactly the time and place it is required in the quantity needed. An example of this is vermicomposted fertilizer, which can be made from city solid waste and delivers exactly the nutrients required by crops over a three-year period via a natural time-release mechanism that occurs as the microorganisms in the fertilizer continue to release the nutrients from the undigested portions of the material.

A Need for More Sustainable-Lean Tools

As well as the above examples, there are myriad opportunities to expand the commercial tools of Lean to become sustainable tools of Lean.

Sustainable Lean will also need to add new tools that commercial Lean does not have because the latter deals with a commercial subset of the issues encompassed by sustainability.

Examples of issues that these new tools need to address must include: closed-loop technology, redesign processes to improve the structural sustainability of communities, reengineering of commercial processes to prevent pollution, and substantive improvements in communication that will both educate and motivate community members to engage more actively in deciding the quality of their future.

Although sustainable Lean may not be recognized as a formal methodology by many, parts of it have been successfully applied when individuals or companies have combined a commercial Lean initiative with a sustainable development initiative and so adapted existing Lean tools to fit their needs. By recognizing sustainable Lean as a structured approach to sustainable development, we can improve the success and speed with which the well-tested principles, rules, and tools of commercial Lean are adapted to sustainable Lean and implemented.

U.S. Environmental Protection Agency and Lean

The Environmental Protection Agency (EPA) commissioned a report by Ross & Associates Environmental Consulting, Ltd. in 2003 to examine the relationship between Lean and the environment (Ross & Associates Environmental Consulting, Ltd. 2003). The report not only supported the use of Lean but also showed how organizations could use their Lean initiatives to further enhance their environmental performance, thereby increasing the speed with which private and governmental organizations are consciously moving toward goals of sustainability.

The report stated that Lean creates a working environment that is highly conducive to the minimization of waste and the prevention of pollution, both of which typically lead to significant environmental benefits. Moreover, it confirmed that Lean usually leads to significant reductions in the use of raw materials, water, and energy, as well as a reduction in waste and a decreased usage of chemicals, in both number and amount.

The report also suggests that further environmental improvement could be achieved by leveraging Lean because it does not explicitly consider potential opportunities for environmental enhancement. In addition, many companies could broaden their definition of waste, thus allowing Lean to become an excellent platform from which to address a product's life cycle and the risk of its introduction into the environment.

The report also mentions that an increasing number of U.S. organizations are supporting Lean. These organizations, like many involved in sustainability, are seeking to eliminate waste and thus reduce the use of raw materials. Because there is little collaboration among organizations, the report found this to present an opportunity for organizations to commence working with one another and thereby increase the social–environmental benefits associated with Lean.

How the Environmental Protection Agency Is Working with Lean

The EPA is partnering with experts in Lean, organizations implementing Lean, state environmental agencies, and other entities to use Lean more effectively. The agency is seeking an increased awareness of Lean's ability to enhance environmental sustainability. The EPA also promotes the sharing of "good practices" for improving the environment, which benefits the implementation of Lean thinking (USEPA 2008c).

The agency is seeking to develop and distribute environment tools integrated with those of Lean to address environmental regulatory issues associated with Lean initiatives. They are also exploring ways that Lean techniques might be used to improve the administrative processes of government related to environmental issues, such as permitting. Coupled with this work, the EPA is documenting environmental success stories related to Lean in a number of industrial sectors, as well as maximizing the environmental benefits of Lean for organizations by developing tools with which to test the overall concept (USEPA 2005).

Summary of the Environmental Protection Agency's Report on the Environmental Benefits of Lean Methodology

The EPA found that *Kaizen* methodology (the process of continuous, incremental improvements that reengineer existing systems to eliminate waste) leads to a culture of continual improvement focused on the elimination of waste, while uncovering hidden waste and waste-generating activities. The results prove to be quick and sustainable, but without significant capital investment (USEPA 2008b).

What we know as 5S activities constitutes a comprehensive methodology for organizing a work space. The 5S comes from the five Japanese words for *separate, sort, sweep, standardize,* and *sustain*. This leads to (1) a decrease in lighting and the use of energy when windows are cleaned and equipment is painted with light colors, (2) the quick detection of spills and leaks, (3) a decreased potential for accidents with clearly marked, obstacle-free thoroughfares, and (4) a reduced contamination of products, resulting in fewer defects, thereby further reducing the use energy and raw materials, which, in turn, becomes a self-reinforcing feedback loop that reduces waste still further by avoiding it in the first place.

The use of Lean methodologies reduces the floor space required for operations and storage, which results in a reduced use of energy. When equipment, parts, and materials are organized and easily found, it reduces surplus materials and prevents the need to dispose of expired chemicals because they are used in a timely manner. A fast turnover of inventory also lessens the need for storage space and facilitates keeping it clean. In addition, cellular manufacturing eliminates overproduction, thereby reducing waste, the use of energy, and raw materials even more. And visual cues can raise awareness of how to handle waste wisely, forewarn of potential hazards in the workplace, and streamline responses to emergencies that might occur.

Lean tools, like error proofing, allow companies to produce more with less, also resulting in fewer defects, which translates to a lower use of energy and resource. Defects are noticed earlier, preventing waste while lowering the use of materials and energy (per unit of production) that comes with Lean "right-sizing of equipment," which reinforces the feedback loop that results in need for less floor space, leading to a decrease in the use of energy, and thus less need to construct new facilities—making it easier to focus on the proper maintenance of equipment and the prevention of pollution.

The Lean just-in-time *Kanban system* eliminates overproduction, thereby reducing waste and the use of energy and raw materials, lowering in-process and post-process inventory, while simultaneously avoiding potential waste from damaged or deteriorated products.

Lean Total Productive Maintenance results in fewer defects, thereby reducing the use of energy and resources, while increasing the longevity of equipment, thus lessening its need for replacement, with all the attending environmental benefits of prolonged life and service. This high quality of maintenance decreases the number and severity of such things as spills and leaks, thus protecting the environment.

Lean uses nature (inherently free of waste) as a design model. For example, (1) right-sized equipment lowers the requirement for materials and energy in production; (2) reducing the complexity of the production process (design for manufacturability) can eliminate or streamline process steps; (3) environmentally sensitive processes can be targeted for elimination, since they are often time-, resource-, and capital-intensive; and (4) less complex product designs can use fewer parts and fewer types of materials, increasing the ease of disassembly and recycling.

Case Studies

The following two examples from the Environmental Protection Agency show environmental benefits that resulted from the implementation of Lean thinking and provide a better understanding of how Lean can directly improve the level of sustainability in an operation:

Apollo Hardwoods Company: Started in 2003, this company provides a unique example of a business enterprise designed and launched with Lean principles in mind. The company manufactures custom, cut-to-size plywood for cabinetry made from fine northwestern Pennsylvania cherry wood. They now use fewer trees and less energy to produce the same amount of product. They have designed equipment that can use smaller pieces of wood, which reduces wood scraps and alleviates the need to harvest large-diameter, mature black cherry trees (USEPA 2008a).

Lockheed Martin: Lockheed Martin reduced hazardous waste resulting in cost savings due to the elimination of the Resource Conservation and Recovery Act permit requirements, which allowed them to reduce the facility by one-third (a reduction of 550,000 square feet), as well as the chemical storage capacity to 2% of its original size (USEPA 2008a).

Lean and Sustainability at Interface, Inc.

Interface, Inc. is an internationally active, Atlanta-based carpet manufacturer. It is a resource-intensive company whose largest divisions are dependent on petroleum. With sales in more than one hundred countries and manufacturing facilities on four continents, the company affects global commerce and ecology.

Ray Anderson, the chairman of Interface, Inc., experienced an epiphany in 1994, after reading *The Ecology of Commerce* by Paul Hawken (1993). That experience motivated him to do something substantial toward achieving sustainability. How does Interface, Inc. define sustainability? "It's more than environmentalism. It's about living and working in ways that don't jeopardize the future of our social, economic and natural resources. In business, sustainability means managing human and natural capital with the same vigor we apply to the management of financial capital. It means widening the scope of our awareness so we can understand fully the 'true cost' of every choice we make" (Interface 1998).

The Interface model for achieving sustainability was built on the foundations of Lean, which the company customized to suit their needs through a program titled Quality Utilizing Employee Suggestions and Teamwork (QUEST). They also expanded the definition of waste to include virtual financial losses, such as the use of nonrenewable energy because it is a financial loss.

Having institutionalized Lean through QUEST, Interface, Inc. has built one of the most successful corporate initiatives on sustainability in the

United States. They continue to fund their initiative from the cash generated through their QUEST program, but the greatest benefit of QUEST is the strong community ethic it has built and continues to maintain. This sense of a common purpose is as valuable as the cost savings provided by eliminating waste (Hay 2006).

Seven Steps Interface, Inc. Uses to Achieve Sustainability (Liker 2004)

The path toward sustainability chosen by Interface, Inc. requires effort on seven fronts:

1. Eliminate waste—Interface's campaign to eliminate the concept of waste, not just incrementally reduce it.
2. Benign emissions—Focuses on the elimination of molecular waste, the emission of which has negative or toxic effects on our natural systems.
3. Renewable energy—The substitution of nonrenewable sources of energy with sustainable ones, while simultaneously reducing the use of energy per se.
4. Closing the loop—Requires redesigning their manufacturing processes and products to create a cyclical flow of materials, a goal toward which the company has made good progress.
5. Resource-efficient transportation—Signifies the exploration of methods to reduce the transportation of molecules (products and people) in favor of moving information. This effort includes the careful location of facilities, well-planned logistics, and the effective use of information technology: videoconferencing, e-mailing, and telecommuting.
6. Sensitivity hookup—The creation of a corporate community within and around Interface that understands the functioning of natural systems and the company's impact on them.
7. Redesign commerce—The redefining of commerce to focus on the delivery of service and value instead of the delivery of materials, coupled with engaging external organizations to create policies and market incentives that encourage sustainable practices.

Participation in Sustainable Lean

Lean thinking demands open communication using all appropriate means at all levels at all times, and thus is the most participative approach to transforming a strictly commercial organization into one focused on social–environmental sustainability. The Kaizen transformation process empowers people to apply their creativity in a carefully structured format without giving them the freedom to damage critical organizational objectives or relationships. (Remember, sustainable Kaizen is ideally a process of continuous, incremental improvements that reengineer existing systems to align them more closely with the biophysical principles that govern nature, with a special focus on the elimination of waste.) This concept of freedom within boundaries is a fundamental commonality of all interactive systems, and thus is essential to every well-run organization or project.

More than any other, this aspect of Lean must be carried over to sustainable Lean, because building a community is perhaps the most critical area of sustainable development. Successful organizations are built on two primary social characteristics: trust and respect. Without these two elements, no organization can thrive for long; it will suffer an early entry into decay and death as it builds structures to compensate for these missing values. Lean builds trust and respect by using a no-blame environment, valuing the participation of every employee, and involving senior management in Kaizen activities at every level. Similarly, sustainable Lean will need to value every contribution, involving the leaders in many activities, where they mix with the community and cultivate an atmosphere of no blame and real appreciation of the best in everyone.

Sustainable Lean will achieve this by using sustainable Kaizens and sustainable visibility to build lasting relationships among all members of a community. (Remember, sustainable visibility ensures the effective communication of critical information among all stakeholders, which translates into the display of clear, understandable metrics in appropriate places, where they are easily accessible to those who wish to use them.) One of the most challenging effects of our headlong rush into a technology-centered society has been—and is—the decay of our culture, whereby much has been lost with respect to human relationships. However, the all-inclusive approach of sustainable-Lean thinking demands that all members are invited to join the process because it requires their involvement, regardless of their beliefs or political motivation. There is, nevertheless, a caveat to such involvement; it must be coupled with wise education to expand social–environmental awareness, while increasing people's proficiency in performing critical tasks.

Appropriate Technology and Its Effect on the Community

Lean thinking is revolutionizing industry by removing the complexity associated with large machines for mass production and unnecessary computer systems. Lean organizations rely more on human-to-human communication within teams than through the use of high technology.

Similarly, sustainable Lean need not employ much technology, which makes it eminently appropriate for nonindustrialized countries, especially Africa, where the culture is suitable for the empowering elements of Lean. In the West, however, the use of personal computers linked to the Internet would be assumed, as would the ability to create an intranet site for communication among project members working in different facilities, perhaps based around the development of a project plan by using a tool like Microsoft Project or Veoproject.com.

Because Lean focuses on the simplest forms of communication, other than technology, it naturally establishes closer bonds among team members. Linking people as they work toward an inspiring common goal is an essential part of Lean sustainability. As the relationships among people grow, they will begin to expand the community element within the group, because sustainability seeks to achieve goals that motivate cooperation.

Budgeting and Financial Considerations

Sustainable Lean, like its predecessor, Lean thinking, requires a small budget. The best Lean initiatives grow organically within an organization, without expensive equipment, consultants, or training materials. The cost of hiring Lean experts to lead the project is always recovered from the reduction in personnel that results from Lean. There is no reason why the same should not be true for initiatives using sustainable Lean.

Some budgetary attention should be given, however, to freeing people from disparate responsibilities. In addition, there must be funding for the appropriate education, acquisition of suitable materials, and a small amount for a consultant who can provide the initial training and guidance while the project is being set up.

Summary and Conclusion

Lean thinking not only offers concrete benefits to sustainable development but also can be used at three different levels: level 1—the elimination of waste, level 2—defining values through Lean thinking, and level 3—switching Lean thinking to sustainable Lean. Each level carries its own degree of modification to Lean, the most modified offering the most benefits in terms of sustainable development.

It is clear from the Lean initiatives of the Environmental Protection Agency and Interface, Inc. that a growing body of knowledge is accumulating around Lean thinking and the resulting sustainability. Lean can be taken to a new level of effectiveness by translating its elements into sustainable Lean. Naturally, sustainable Lean subsumes commercial Lean because economics is a subset of sustainability and commercial Lean works primarily in the world of economics.

Since the business sector is already poised for a rapid transformation toward sustainable practices, it is clear that the introduction of a familiar methodology, Lean thinking, in the new guise of sustainable Lean is both appropriate and timely. It will also be effective, however, when used by communities endeavoring to become sustainable.

There is no doubt that we can look forward to the development of sustainable Lean, under this or some other name, as the next logical step in our progression toward global sustainability.

References

3M Corporation. 2007. *3M Lean Six Sigma*. St. Paul, MN: 3M Corporation.

3M Corporation. 2008. 3P examples. http://solutions.3m.com/wps/portal/3M/en_US/global/sustainability/management/pollution-prevention-pays/3p-examples/ (accessed April 11, 2008).

Edwards, Andres R. 2005. *The sustainability revolution: Portrait of a paradigm shift*. Gabriola, British Columbia: New Society Publishers.

Hawken, Paul. 1993. *The ecology of commerce: A declaration of sustainability*. 1st ed. New York: HarperCollins Publishers.

Hay, Buddy. Vice president of sustainable operations, Interface, Inc. August 9, 2006.

Interface Inc. 1998. Company information: Sustainable enterprise introduction. http://www.interfaceinc.com/eur/company/sustainability/frontpage.asp (accessed April 7, 2008).

Liker, J.K. 2004. *The Toyota way: 14 Management principles from the world's greatest manufacturer*. New York: McGraw-Hill.

Ross & Associates Environmental Consulting, Ltd. 2003. *Lean manufacturing and the environment: Research on advanced manufacturing systems and the environment and recommendations for leveraging better environmental performance*. EPA100-R-03-005. Washington, DC: USEPA.

USEPA. 2005. *Lean manufacturing & the environment: Fact sheet*. EPA-231-F-05-003.

USEPA. 2008a. Case studies and best practices/Lean and the environment | EPA. http://www.epa.gov/NCEI/lean/studies/index.htm (accessed April 7, 2008).

USEPA. 2008b. Lean thinking and methods/Lean and the environment/EPA. http://www.epa.gov/NCEI/lean/thinking/index.htm (accessed April 7, 2008).

USEPA. 2008c. EPA lean activities/Lean and the environment/EPA. http://www.epa.gov/lean/activity.htm (accessed April 7, 2008).

Womack, James P., and Daniel T. Jones. 1996. *Lean thinking: Banish waste and create wealth in your corporation*. New York: Simon & Schuster.

6

Working Together to Sustain Local Economy, Environment, and Community: The University of Minnesota Regional Sustainable Development Partnerships

Kathy Draeger, Linda Ulland, Okechukwu Ukaga, and Michael Reichenbach

CONTENTS

> A grand experiment has been unfolding across the Minnesota landscape ... the University of Minnesota Regional Sustainable Development Partnerships.... This is an experiment borne of faculty idealism, Minnesotan agrarian populism, and legislative funding to create citizen-driven University partnerships fostering sustainable development for regional resilience.
>
> **(University of Minnesota Regional Sustainable Development Partnerships 2007, p. 6)**

Kent Scheer, an artisan toy maker with a small acreage in central Minnesota, had an idea. Kent wanted to do something positive for the environment and perhaps earn a bit of money as well. He had heard of carbon sequestration and selling carbon credits, but needed to know more.

Perhaps this was an opportunity for him, and for other small landowners, to practice environmental stewardship and add to their incomes.

Kent approached the University of Minnesota's Central Region Sustainable Development Partnership with his idea. He was looking for information and help with research. The fourteen citizens and university representatives that make up the Central Region Partnership's board of directors discussed how Kent's idea might fit their mission and goals. Will there be active citizen leadership? Does the project support sustainable systems related to agriculture, natural resources, tourism, or energy? How does this further community–university partnership?

Meanwhile, on the St. Paul campus of the University of Minnesota, faculty and students in the Water Resources Center and in the Center for Integrated Natural Resources and Agriculture Management (CINRAM) were researching the biophysical, economic, and market aspects of terrestrial carbon sequestration in Minnesota. Since 2005, the researchers had been assembling scientific information estimating carbon sequestration capacities in Minnesota, the costs and benefits of carbon-enhancing practices, and were now seeking demonstration projects for the most promising strategies.

The Central Region Sustainable Development Partnership board saw great potential in bringing together their local community members with these university resources, and voted to support and fund the project. The subsequent support by university faculty, students, and the staff of the Central Region Sustainable Development Partnership moved Kent's ideas forward. One year later, the *Landowner's Guide to Carbon Sequestration* was published. It provides information to small landowners on the opportunities to sell carbon credits, and the processes required to participate in such programs (Current et al. 2007). The guide is used by the University of Minnesota Extension faculty to conduct educational and training sessions for landowners in rural Minnesota. As a result of the Partnership's work, Kent enrolled his lands in a carbon credit program with the Chicago Climate Exchange.

This vignette highlights one of the more than four hundred projects of the University of Minnesota's Regional Sustainable Development Partnerships and highlights the essence of the partnership:

- Building effective relationships among citizens, communities, and the university
- Promoting active citizen leadership
- Working together to achieve regional sustainability by investing in research, education, and outreach projects

The compact between society and its public institutions of higher education traditionally entails that the public university educate its citizenry and conduct research and outreach to enhance civic society and quality of life. Land-grant universities are charged with serving citizens in each state through education, research, and outreach. The University of Minnesota Regional Sustainable Development Partnerships serve as a framework from which to build on the compact between society and public institutions of higher education. The Partnerships can keep the compact vital and help create a sustainable society.

The Regional Partnerships officially began in 1997, when the Minnesota legislature approved $1.2 million per biennium for three pilot Regional Partnerships. In 1999, the legislature increased the appropriation to $2.4 million per biennium to support five Regional Partnerships and a statewide body to coordinate efforts among the regions and the university. But the idea for the Regional Partnerships started years earlier.

The concept for the Regional Partnerships program began five years before the legislature appropriated funding to implement the program. Dr. Don Wyse, then executive director of the Minnesota Institute for Sustainable Agriculture (MISA), along with other university faculty and citizens, began working to make the university more responsive to the citizens of the state (Peters et al. 2000, 87–96). MISA, itself a partnership between the university's College of Agriculture, Food and Environmental Sciences and a coalition of individuals and nonprofit organizations, provided a model for university engagement. Dr. Wyse sought to have this model regionally based and directed by both community and university board members.

In the early to mid-1990s, there was a confluence of local and statewide factors:

- Rural communities in the 1990s were experiencing financial difficulties due to a variety of economic, political, environmental, and cultural conditions.
- Interest in the concept of sustainable development was taking hold, and was reflected in a 1996 Minnesota state statute that charged the state to follow a sustainable development path.
- A new understanding of the civic nature of the land-grant mission reemerged to involve citizens as full participants in shaping and conducting the university's research, education, and outreach work (Peters and Lehman 2005).

In 1993, in the central region of the state, these factors were affecting research and educational opportunities at what is now the Central Lakes Agriculture Center outside Staples, Minnesota. The center had served as an applied research and demonstration site for irrigation practices for area farmers and University of Minnesota researchers. Dr. Wyse was among the university faculty leading agricultural research at the center when it was challenged to develop a new mission and purpose. Recognizing the need for broad community input, a group of citizens organized local focus groups to identify community priorities and issues related to agriculture, natural resources, and the environment. This effort resulted in the report *Concerns from the People of Central Minnesota* (Thorson 1994).

Between 1994 and 1997, MISA staff and board led by Wyse, with input from local citizens, legislators, students, and faculty, developed the concept into a specific proposal that was submitted for legislative funding in 1997. Once the funding was approved, the task of implementing the program began. A statewide task force of citizens and university members developed operating principles and guidelines and refined the regional boundaries for the Partnerships (University of Minnesota Regional Partnerships Guidelines & Operating Principles, 2005).

A task force was established in each of the three pilot regions (central, northeast, and southeast) to initiate a Sustainable Development Partnership program in their region. When planning for implementation began in each of the regions, a number of issues arose. The most challenging was a discussion of the nature and degree of authority shared between the university and the regional boards. Each region was concerned with the amount of power, influence, and control regional boards and the university would have over the Partnership's work. The discussion resulted in an agreement to focus on the notion of partnership, wherein influence was shared. The regional boards, comprised of both citizens and university personnel, determined project priorities and funding.

More than ten years of community engagement and faculty advocacy through the University of Minnesota Regional Partnership has evolved into a program to establish partnerships and networks among the citizens of Minnesota, regional and state agencies, and the university in a citizen-driven process that identifies local issues and seeks solutions. As a result, the Regional Partnerships have increased community awareness, active involvement in, and access to research, as well as education and outreach by the university. We believe this framework can be adapted nationally and internationally.

The Sustainable Development Partnerships

Universities as institutions can engage society and address some of the critical environmental, social, and economic issues of our times. Dr. William

Clark, the Harvey Brooks Professor of International Science, Public Policy and Human Development at Harvard University, was in Minnesota in September 2007 to evaluate work of the University's Ecosystems Science and Sustainability Initiative (Sustainability Initiative), of which the Regional Partnerships are a collaborator.

Dr. Clark, enthusiastic about the Regional Partnership as a model and the work with the Sustainability Initiative, said:

> This effort is incredibly innovative and cutting edge ... linking science in the service of community. This is not disinterested science—this is getting at sustainability through engagement of expertise like in fisheries and soils with the people who live it, the farmers and county commissioners. This blends action and knowledge more closely and brings in the values and the politics. Reciprocal respect—that is the name of the sustainability game ... this is amazingly exciting. Looking downstream 10 years the leading universities will be the ones who know they serve their state and harness the forefront of research to bring it to a mutually respected place that has been grappling with community well-being and sustainability. (Clark 2007)

The university and communities need each other to find a way to a sustainable future. The Regional Partnerships provide an adaptable model of a "mutually respected place," where community and university come together for the well-being and sustainability of a thriving Minnesota. This is consistent with the university's avowed commitment to public engagement and civic responsibility.

> As a signatory of the Presidents' Fourth of July Declaration on the Civic Responsibility of Higher Education, the University of Minnesota is committed to reinvigorating its public purpose and civic mission, joining a national movement of over 400 colleges and universities devoted to the task of strengthening higher education institutions' role as "vital agents and architects of a flourishing democracy." (Peters et al. 2000, 94)

The Partnerships support sustainable development in Greater Minnesota by facilitating public engagement and utilizing university resources to address community-identified research, education, and outreach. It is based on the concept of connecting citizens to their land-grant university, which is as old as the land-grant philosophy itself. Brought into being by Senator Justin Morill of Vermont and President Abraham Lincoln, at the height of America's Civil War, the land-grant philosophy mandates both citizens' access to education and universities' service to communities (Christy and Williamson 1992, 166; McLaurin et al. 2000, 207).

The World Council on Environment and Development (WCED) further catalyzed global conversation about sustainable development, which it defined in *Our Common Future* as development that meets the needs of the present without compromising the ability of future generations to meet their own needs (World Commission on Environment and Development 1987, 383). In 1996, the Minnesota legislature adopted the above WCED's definition of sustainable development and noted that sustainable development "maintains and enhances economic opportunity and community wellbeing while protecting and restoring the natural environment upon which people and economies depend." Realizing this vision of sustainable development is a challenging task that requires serious efforts, including not only an appropriate blend of organizing, research, education, outreach, and policy making, but also creative spaces and new institutions that support and facilitate innovation and positive change (Peters et al. 2000, 87–96).

Purpose and Approach

In collaboration with the university faculty and students, citizen leaders in local communities work through regional boards to develop and support a variety of innovative research, education, and public engagement projects that facilitate local sustainable development in agriculture, natural resources, tourism, energy, and other areas. In so doing, the Partnerships help address perceived university challenges, such as erosion of support for investment in higher education and "public interest" scholarship. By providing safe zones for complex community conversations about important needs and issues, as well as where to find necessary funding and staff support to move projects forward, the Partnerships play a unique and critical role in helping the university and the community work together to meet the needs of both community and university stakeholders.

Principles and Goals

The three bedrock principles that guide the Regional Partnerships are to (1) foster community and university partnerships, (2) promote citizen engagement, and (3) facilitate sustainable development. Working in the context of these principles, regional boards made up of both university personnel and local citizens identify issues, needs, and opportunities for sustainable development at both the local and regional levels; work directly with all stakeholders to generate ideas, formulate goals, and make investments in selected research, demonstration, education, and outreach projects; monitor and evaluate funded projects; and share results in their region, across the state and beyond.

Sustainable Development

The Regional Partnership boards' understanding of sustainability has grown over time, and some report that, as a result of this enhanced understanding, their projects are better examples of sustainable development than what they did at the beginning. The Regional Partnerships create an understanding of sustainability by the projects they fund. This transcends the endless, abstract debates that can occur over the definition of sustainable development. A University of Minnesota Regional Partnership board member as part of a 2006 focus group stated:

> I never came up with a definition of sustainability that I bought into.... I can deal with "is this sustainable" or "what could make it sustainable," 10 times better than I could deal with "this is what sustainability means." And so it is an application of very general sustainable values and saying is this something that we can say in our hearts that it is sustainable? And that is why we can support it. (Krueger et al. 2006)

Structure and Governance

The activities of each Regional Partnership are directed and managed by an executive director and a fifteen- to twenty-member board comprised of two-third citizens from the region and one-third University of Minnesota faculty. In addition, cross-regional communication, learning, and collaboration are facilitated by a statewide director and the statewide coordinating committee (SCC), which consists of staff from all Regional Partnerships, two individuals representing each partnership, two at-large citizen members, and one representative from each of the partnership colleges. A rotating regional board and SCC membership is encouraged. Diversity and balance are sought in the regional boards in terms of members' geographic location, gender, ethnicity, age, professional experience, and interest. Each board creates a vision for regional sustainable development in their own area, and then works to facilitate selected community-based projects that can be implemented, in partnership with the university, toward that vision.

Accomplishments

> Partnerships and collaborative initiatives leverage opportunities to bring University resources directly into communities to address complex challenges. These partnerships and collaborations also bring real-world problems to the attention of University researchers, sparking new and important directions of inquiry. The Regional Sustainable Development Partnerships bring citizens and university

> faculty and students together to strengthen rural Minnesota. Five Regional Partnerships fund research, education, and public service projects to strengthen the long-term sustainability of their regions' natural resource based initiatives.
>
> **—Dr. Robert Bruininks, President, University of Minnesota**

The Regional Partnerships have accomplished much since the program became fully staffed in the year 2000. They have created an organizational structure, goals, boards, and systems for implementing the model. The Regional Partnerships, as of December 2008, have funded over four hundred projects and supported more projects in ways other than funding. Interviews of stakeholders and university faculty, in 2006, show that people think the Partnerships have been successful at promoting active citizen leadership; advancing the understanding of and achievement in regional sustainability; and strengthening relationships between citizens, communities, and their university.

Active Citizen Leadership

Findings from the 2006 focus groups conducted as part of the University of Minnesota Regional Partnerships evaluation report show that the Partnerships are providing safe zones for community discussion and initiatives (Regional Sustainable Development Partnerships 2007, 1–35). Each Partnership has deeply committed citizen leadership and partners. There is a network of enthusiastic, hardworking people within each of these regions that is committed to Regional Partnerships and their projects. Board members work well together. They feel free to share diverse opinions, are respectful of others, and enjoy working together. The Partnerships are also learning more about how to foster citizen-driven projects.

Further, the focus groups show that people believe the Regional Partnerships are a catalyst in the revival of rural areas and provide connection to the university to help revitalize Greater Minnesota. And they already see and appreciate successes and outcomes of the partnerships. The following quotations from two stakeholders illustrate the above point:

> I have nothing but compliments for the idea [Regional Partnerships].... I think the concept is great, that citizens have to be involved and ask their University to help solve their problems. And the University, as a result of that small amount of funding, entices the expertise of people from the University to help the people out.
>
> **—University of Minnesota Regional Partnership community partner, 2006 focus group (Krueger et al. 2006)**

> I live in desperate need of hope. There is hardly any hope to be seen no matter where I look. In every one of these [Partnership] projects there is a tiny sign of hope.... All is not lost. There are people doing good work.
>
> **—University of Minnesota Regional Partnership community partner, 2006 focus group (Krueger et al. 2006)**

From the community perspective, the Partnerships have, with relatively small budgets, been successful in promoting sustainable development, fostering vital university and community relationships, and facilitating active citizen engagement and leadership. As a result, various communities received:

- Unusual access to the university
- Opportunities to commission research and other initiatives
- Tangible products, including research findings, databases, and community plans
- A lot of work for relatively small amounts of money
- Increased capacity to meet other challenges

University of Minnesota Involvement

Over nine hundred faculty and three thousand students have engaged with communities and other stakeholders across Minnesota through Regional Partnerships projects. Board members believe their relationships with the university have evolved to be more cooperative and less contentious. Initially, some boards found it difficult to balance the goals of active citizen leadership and university involvement in projects. Some projects had extensive citizen leadership but little university involvement. Other projects had extensive university involvement but little citizen leadership. The Regional Partnerships now have better strategies for accessing and involving university faculty, staff, and students. They have also learned more about what works and what does not. Further, four of the Regional Partnerships are in the areas with University of Minnesota coordinate campuses (Northeast, Northwest, Southeast, and West Central) and have created strong partnerships with their local campuses. Notably, the Partnerships have evolved to be a better consumer of university resources, and university people have learned more about how to work with the Partnerships and what type of research is useful to communities. Thus, relationships have improved, resulting in a very effective program, staff, boards, and projects.

> The University involvement has taken a big leap.... We have evolved to be more competent in terms of what we want to do.... We are not

> in a narrow little focus any more. It has come a long way. You really have an open door [to the university now].... We can do anything you want.... Also changes have taken place at the University as well.... Those [university] folks have learned along the way with us about what works.
>
> **—University of Minnesota Regional Partnership board member, 2006 focus group (Krueger et al. 2006)**

Feedback from University of Minnesota researchers and faculty members who have been involved in one or more partnership projects reveal that their university work (teaching, research, and public engagement) has become more applied and relevant as a result of working with the Regional Sustainable Development Partnerships. University of Minnesota faculty who participated in focus groups conducted as part of a recent evaluation of this initiative noted several characteristics that, together, make this model unique and effective (Regional Sustainable Development Partnerships 2007, 1–35).

The program is a grassroots participatory public engagement effort. Citizens identify the issues that are important to them, make the decisions about what will be funded, and have power to collaborate with other organizations toward mutual goals. As a result, citizens have a direct influence on the teaching, research, and outreach agenda of the university. Further, the Regional Sustainable Development Partnerships provide needed infrastructure and a stable platform for civic engagement. A university of Minnesota faculty member commented as follows:

> My research has become more "applied" rather than theory, which has led to testifying at the Minnesota Legislature on behalf of "rural" conditions. I found that most policies do not recognize the rural difference or challenges as most policies are developed with Metro lenses. Healthcare coverage was especially an eye opener ... in the West Central Minnesota area, 75% of business employ less than 10 people. Yet, in waiver and tax breaks are given only to those employing 50 or more ... leaving out much of rural Minnesota. (Regional Sustainable Development Partnerships, p. 14)

Lessons Learned

The role of regional executive director is exceedingly challenging and requires strong social and managerial skills, coupled with a familiarity and understanding of both the community and the university. Because the executive director's role is critical to the success of the Regional Partnerships, careful thought must be given to recruitment, training, and then providing regular and meaningful feedback and support to these individuals.

There is need to create and maintain the right balance between community and university interests, cultures, and timelines. For instance, communities like things to happen quickly, while faculty often need long lead times. Communities also like to solve problems that may or may not be neatly packaged, while universities like to teach or research well-defined subjects.

Successful projects are usually those that were proposed by a community member rather than a faculty member; fit the university's mission and method of operation; had good faculty leadership, plan of work, adequate funding, and timeline; and were brokered by the executive director (who helped translate between the university and community stakeholders and negotiated what was doable within a certain budget and timeline).

The Partnerships serve as a framework for engagement and scholarship. Ernest Boyer (1990) described methods to make the university's work more relevant through his *Scholarship Reconsidered: The Priorities of the Professoriate*. Boyer called for expanding the concept of scholarship within the university, broadening its definition from research (what Boyer called discovery) to include teaching, integration, and application. The Partnerships can help university faculty develop rigorous programs in all elements of scholarship by engaging citizens and faculty in ways that create new insights, explore various ways of being, learning, and doing, as well as making the work richer, more meaningful, and productive.

Summary and Conclusions

The University of Minnesota Regional Sustainable Development Partnerships support sustainable development in Greater Minnesota by facilitating public engagement and utilizing university resources to address community-identified research, education, and outreach needs. Toward this end, the Regional Partnerships help develop and support projects that build participatory relations between citizens, communities, and their university; promote active citizenship in strengthening the future and long-term social, economic, and environmental health of Greater Minnesota; and invest in research, education, and outreach projects that advance the understanding and achievement of regional sustainability.

The Regional Partnerships represent a unique and new approach to university public engagement. Contrary to the normal top-down, university-led, expert-driven approach, the Regional Partnerships provide a creative space that belongs to neither the community nor the university, but instead to both community and university stakeholders who work together in a participatory fashion to make situation statements, generate project ideas, define needs, and envision solutions, resources, and partnerships needed. Out of this participatory interaction emerge specific research, education, or outreach ideas that can evolve into projects with ownership by all community and university stakeholders. Perhaps most important is the sense of hope having the University of Minnesota actively engaged in community-identified projects brings to all the regions throughout Minnesota.

How has the Partnership model served Kent Scheer and his community? Beyond offering a value-added opportunity for small landowners and encouraging environmental stewardship, it has helped Kent develop leadership skills; he now refers to himself as a volunteer public servant. It has connected the university and the community. The community now looks to the university as *their* university and seeks opportunities to work together in other ways.

References

Boyer, Ernest L. 1990. *Scholarship reconsidered: Priorities of the professoriate*. Princeton, NJ: Carnegie Foundation for the Advancement of Teaching.

Christy, Ralph D., and Lionel Williamson. 1992. *A century of service: Land-grant colleges and universities, 1890–1990*. New Brunswick, NJ: Transaction Publishers.

Clark, William. 2007. Harvey Brooks professor of international science, public policy and human development at Harvard University. Kathy Draeger, personal communication.

Current, Dean, Kent Scheer, John Harting, Diomy Zamora, and Linda Ulland. 2007. *A landowner's guide to carbon sequestration credits: In association with the Commonwealth Project*. St. Paul: University of Minnesota, Center for Integrated Natural Resource and Agricultural Management and Central Minnesota Regional Sustainable Development Partnership.

Krueger, Richard, Mary Anne Casey, and Randi Nelson. 2006. Summary of focus groups for the Regional Sustainable Development Partnerships. Unpublished report submitted to the University of Minnesota Regional Sustainable Development Partnerships.

McLaurin, Sylvia, University of Georgia, and Kellogg Commission on the Future of State and Land-Grant Universities. 2000. *Transitions into the 21st century: Past, present, and future*. Athens: Public Service and Outreach, University of Georgia.

Peters, Scott J., and Karen Lehman. 2005. Organizing for public scholarship in SW Minnesota. In *Engaging campus and community: The practice of public scholarship in the American land-grant university system*, ed. Scott Peters, 67–117. 1st ed. Dayton, OH: Kettering Foundation.

Peters, S.J., K.P. Lehman, J. Ristau, O. Ukaga, and D. Wyse. 2000. The promise and challenges of the regional agricultural and natural resources sustainable development partnerships. In *Just in time research: Resilient communities*, 87–96. Vol. 2. St. Paul: Hubert H. Humphrey Institute of Public Affairs; University of Minnesota Extension Service.

Thorson, Eloise. 1994. *Concerns from the people of central Minnesota: Report to Central Minnesota Economic Development, Research and Education Center, a summary of six focus group sessions*. Staples, MN: Brainerd Staples Regional Technical College.

University of Minnesota Regional Sustainable Development Partnerships. 2005. Guidelines & Operating Principles. Working document. www.regionalpartnerships.umn.edu/public/Guidelines. pdf

University of Minnesota Regional Sustainable Development Partnerships. 2007. *The university and the community: Creating a sustainable future through civic engagement*. St. Paul: University of Minnesota.

World Commission on Environment and Development. 1987. *Our common future*. Oxford: Oxford University Press.

7

The Holistic Management Framework: Ensuring Social, Environmental, and Economically Sound Development*

Allan Savory and Jody Butterfield

CONTENTS

* Holistic Management® is a registered trademark of Holistic Management International, founded by Allan Savory and Jody Butterfield.

The Savory Institute works with land stewards on four continents to heal damaged land, improve biodiversity and food production, and reverse desertification. Thirty million acres worldwide are currently under Holistic Management. This chapter presents background about the process of Holistic Management and two case studies from Zimbabwe, Africa, where the basic concepts of the Holistic Management framework were initially formulated.

As a youngster, Allan Savory's only aim in life was to live forever in the wildest African bush. Though he eventually had that opportunity in what is today Zambia and Zimbabwe, he ended up forsaking it in order to work toward saving the wildlife that was his reason for being in the bush. Even in the wildest areas, the land was deteriorating, in fact, turning to desert, rendering it ever less able to support life of any kind, and he was determined to find a way to reverse this process.

That quest to save wildlife compelled him to work with people who had been caretakers of the land for generations. Soon thereafter, he began working with those who were advising the caretakers, and eventually with many others as a member of parliament, where he dealt with land management at the policy level.

From these experiences, as well as extensive reading of the research, he learned that the remorseless spread of deserts and the human impoverishment and violence that always resulted were fundamentally related to the way people were *making* management decisions, whether those people lived and worked on the land or not. Hence, Savory and his colleagues developed the Holistic Management framework as a way forward for making development more sustainable.

Holistic Management is a framework for decision making and management that enables people to satisfy their immediate needs without jeopardizing their well-being, or that of future generations. It does this by ensuring that actions taken in meeting those needs enhance the well-being of the environment that sustains them now and in the future. Toward that end, a number of procedures and guidelines for planning and management are incorporated in the framework.

Overview

Sustainable development rests on a sustainable agriculture, which is impossible to achieve if land is degrading. Today, the symptoms resulting from land degradation, including desertification, are exploding exponentially: increasing frequency and severity of droughts and floods; massive invasions of noxious plants and insects; poverty; violence, including abuse of women and children as populations exceed declining resources; genocide; and, of course, migration from rural areas to cities. These symptoms are common in nonindustrialized and industrialized countries alike, including the United States. Because the only form of wealth that can sustain a nation is derived from the photosynthetic process, in which green plants convert solar energy to usable forms, including food, any attempt to sustain development that does not address land degradation will always be short-lived. The Holistic Management framework takes this fact as a basic premise.

Each of us, no matter what path we have chosen to follow in life, makes decisions that in one way or another affect the health of our environment and the quality of other people's lives. Given a simple technique for determining what that impact might be goes a long way toward ensuring a more sustainable, life-enhancing outcome.

In brief, this Holistic Management begins by defining the entity being considered in terms of the people responsible for its management and the resources available to them. These people then form a "holisticgoal" that describes (1) the quality of life they collectively seek, (2) what they have to produce to create that quality of life, and (3) the resource base they depend upon *as it will have to be*, far into the future, to sustain what they must produce to create the quality of life they desire.

All the decisions made in addressing problems or opportunities that arise along the way are based on familiar criteria—expert advice, cost, peer pressure, research results, and so on. In addition, however, the people ask seven simple questions to ensure their decisions are socially, environmentally, and economically sound *and* will lead them toward the holisticgoal. In other words, any action taken to deal with a problem, to reach an objective, or to meet a basic need must accomplish what is required and simultaneously enhance progress toward the holisticgoal. To ensure that it does, a monitoring feedback loop is established to allow immediate course corrections, if required.

Key Principles of the Holistic Management Framework

Holistic Principle 1: Addressing the Root Cause of Environmental Deterioration Is Essential

People often struggle to accept that the root cause of environmental deterioration is allied to decision making. Instead, they blame the deterioration on overgrazing, overpopulation, communal rather than private land holding, poverty, greed, ignorance, corruption, and a host of other factors. For example, Figure 7.1 depicts a fence going through the middle of seriously degraded land in New Mexico. On one side of the fence, Navajo Indians have been grazing livestock for several centuries and are blamed for causing the deterioration through overstocking, overgrazing, poverty, ignorance, communal tenure, or a combination of these. On the other side of the fence, the U.S. National Park Service has applied its "best management practices," meaning that no livestock has been present for over seventy years and hundreds of thousands of dollars have been spent on soil conservation measures. Yet no one can tell the difference after so many years. The land is turning to desert on both sides of the fence, despite opposing management practices.

Both the Navajos and the National Park Service were basing their management decisions on goals and objectives, many of which were probably met in the short term, but not without damaging the social, economic, and environmental base in the long term. And they are not alone. All tool-using animals do the same thing—be it an otter using a stone to break shells, a vulture using a stone to break eggs, an ape using a stick or grass stem to catch ants, or a scientific team using rather more sophisticated

FIGURE 7.1
Livestock were removed more than seventy years ago from the land on the right, which is managed by the U.S. National Park Service. Navajo stockmen have grazed their livestock continuously on the land left of the fence.

tools to put a man on the moon. This pursuit of objectives, at the expense of the greater whole, appears to be programmed into our genetics.

The "tools" humans use in manipulating their environment boil down to fire (deliberately burning vegetation), technology (any human-created artifact), and more recently, rest (nondisturbance). There is a fourth tool, living organisms (plants or insects), that some societies have used for centuries to manage crop and garden pests, and increasingly modern farmers use too. The challenge with these few tools is that none of them can assist in promoting the decay of vegetation in those environments where long periods of low or no humidity, low microorganism populations, and the absence of large herbivores prevent decay from occurring. These seasonally humid environments cover about two-thirds of the Earth's land surface.

The rank vegetation that remains at growing season's end, if left undisturbed (or rested) for one or more seasons, begins to oxidize and then slowly breaks down with weathering. As a result, the dead leaves choke out new growth on perennial grasses and grasslands begin to die out. They are replaced by woody vegetation or bare ground and forbs, depending on the rainfall. No technology exists that can promote rapid decay over the billions of acres that require it. The only remaining tool—fire—is the one we generally use by default, as it is the only tool we have to clear away the decadent, oxidizing vegetation to make way for the new. But fire doesn't promote biological decay; it promotes rapid oxidation and it always results in soil exposure. Exposed soil in turn promotes desertification.

Holistic Principle 2: Grazing Animals Are Essential for Maintaining Biological Decay

In seasonally humid environments, it is essential to add two additional tools to the tool bag—grazing and "animal impact" (trampling, dunging, urinating, digging, etc.). Only large herbivores in symbiosis with large populations of microorganisms in their digestive tracts can maintain biological decay of annually dying plant material—mainly grasses. In the past, this service was performed in these environments by vast numbers of grazing ungulates. When too few herbivores are present to maintain biological decay, dead vegetation then breaks down very slowly through oxidation (plants turn gray) or weathering (Gordon and Duncan 1988, 54–59; Hobbs and Huenneke 1992, 324–37; McNaughton and Banyikwa 1997, 1798; Pickett and White 1985, 472). Cities in seasonally humid environments cannot be sustained by the lands surrounding them, which in nearly all cases are desertifying to one degree or another, other than through the use of properly managed livestock.

Holistic Principle 3: Timing, Not Numbers, Produces Overgrazing

Overgrazing is not a function of animal numbers, but rather of time of exposure and reexposure of plants to animals (Voisin 1960, 341; 1962, 85; 1988, 341). When livestock are used to restore land suffering from desertification, their movements must be planned to ensure they don't stay in one place too long or return to it too soon. One animal that stays too long or returns too soon will overgraze plants. But thousands of animals can occupy an area for three days or less and not overgraze plants, provided they don't return before plants and soils have recovered—anywhere from about one to six months or more.

The holistic grazing planning procedure has been developed over the last fifty years to address the timing issue, as well as a great deal of complexity—weather, droughts, wildlife needs, habitat creation, and much more. Those designing grazing systems have through the ages tried to sidestep this complexity. In principle, complexity has to be embraced. Sustainability depends on it.

Holistic Principle 4: Assume You Are Wrong When a Decision Affects the Environment

Even people who have researched and tested their decisions and plans carefully must still assume they are wrong, due to the complexity that exists in any ecosystem. On this assumption, it is necessary to monitor the implementation of decisions and plans for the earliest signs of change, so corrections can be made before damage is done. It is folly to assume you are right when ecosystems are so complex and nature is constantly altering circumstances. What's more, many problems only emerge well after a plan is implemented and far from the site of the original action taken.

For example, in the mid-1950s, the U.S. Army Corps of Engineers built a dam across the Columbia River, which forms the border between Washington and Oregon in the northwestern United States. When the "pool" behind the completed dam was filled, the high water inundated, and effectively destroyed, the habitat of the Columbian white-tailed deer.

Some years later, the deer was declared an endangered species, and the Corps was ordered to mitigate its habitat by creating suitable habitat to replace that which was lost when the dam was put into operation. What no one considered, however, was that artificially creating habitat for the Columbian white-tailed deer (a positive trade-off for the deer) caused more than three hundred other species of plants and animals to lose their habitat (a negative trade-off for all of them).

That we often neither understand nor anticipate trade-offs lies in the fact that what informs our perception is most often an appearance of something, rather than its underpinning reality, because as American author Anaïs Nin observed, "We don't see things as they are. We see things as we are."

Holistic Principle 5: Manage for the Health of the Whole Biological Community, Rather Than Individual Species

No amount of captive breeding, plantings, culling of predators, poisoning of competitors, or other narrowly focused actions will bring back a plant or animal that has lost its niche or habitat. The surest way to truly save any species, other than short-term efforts involving slow-breeding mammals, is to manage for the health of the whole biological community.

A case in point is the near disappearance of bushbuck, reedbuck, and duiker (a small antelope) on a ranch in Zimbabwe used as a case study below. All three of these antelope species were causing great concern as their numbers had dwindled to dangerously low levels due to habitat damage and the buildup of a very large population of baboons, which prey on young antelope. By virtually ignoring the plight of the at-risk populations and concentrating instead on increasing ground cover, and thus enhancing the effectiveness of the water cycle and the growth of vegetation generally, numbers of all three species recovered over about an eight-year period. They have in fact increased to such an extent that these species are once again being hunted.

Holistic Principle 6: The Only Form of Wealth That Can Sustain a Community or Nation Is Derived from the Photosynthetic Process

Without agriculture in its broadest sense, it is simply not possible to enjoy the fruits of civilization because there would be none of the many benefits we take for granted today. Agriculture—cropping, pasturing, forestry, and fisheries—is totally dependent upon the conversion of sunlight to ingestible energy in the form of green plants. On land, these green plants are likewise dependent on vast populations of microorganisms in healthy *living* soils. Even today's addictive use of energy from coal, oil, gas, and the subsequent fertilizers pumped into depleted soils to boost the production of green vegetation is based on the sun's energy converted to a usable form by green plants millions of years ago.

Development Principles

While the above principles help define Holistic Management, there are four additional principles that guide us when employing the framework in a rural development context.

Development Principle 1: Much of the Violence Communities Face Is Due to Burgeoning Human Populations and Diminishing Resources

Few would argue this point. But many would argue with the next point: in the two-thirds of the world that is characterized by seasonal humidity—the world's rangelands, grasslands, and savannas—the diminishing resources are not the result of overexploitation, but just the opposite. The problem is underuse—too few animals and, in some cases, few people on the land. The case studies described later in this chapter help illustrate why this is so. The opposite holds true in the perennially humid environments as well as the world's oceans, where resource scarcity is a result of overexploitation, combined with pollution from fossil fuel derivatives.

Development Principle 2: Population Reduction Is Directly Dependent on the Education and Empowerment of Women

Until women are educated and empowered as equals to men, family size will continue to expand as long as food and medical aid are rendered to families. In some African countries, aid organizations commonly promote three development interventions:

- Provide new water from wells (boreholes) or dam.
- Feed and provide health care for starving or malnourished mothers and children
- Improve the local economy through such measures as capitalization through micro-lending, education, assistance with improving infrastructure, and marketing

Although these interventions are well meaning and needed, we are simply chasing objectives that may well succeed in providing immediate relief, but only exacerbate the problems in the longer term. Providing new sources of water merely expands the desert, as countless satellite photos illustrate, thus increasing the scarcity of both forage and water. As long as women are subservient, feeding starving women and children while improving health merely results in larger families and donor dependency. As long as women remain subservient, improving the local economy merely results in larger families.

Experience in northern Kenya, where desertification and the consequent violence are exploding uncontrollably, confirms that violence is often the end result of population growth and diminishing resources, both of which can be exacerbated by external aid. Personal communication with a U.S. Agency for International Development (USAID) worker confirmed the above point. She said they had been feeding people in the region for forty years, and there were now five times as many people killing each other.

Therefore, when training people to manage their lives and resources holistically, three objectives are employed: (1) women should be brought into project leadership positions, (2) women should be made aware of their legal rights, and (3) women should receive training, along with the men, in gender empowerment.

Development Principle 3: The People in a Community Are the Only Experts on That Community

Every community is unique. Also, each community changes and is different every year. This is in part because the people, whose values and aspirations will govern what they choose to do, are being born, marrying, divorcing, moving, and dying. Meanwhile, the economy and the markets within which they operate are ever changing, which, along with the variability of weather and management practices, ensures that no piece of their land will ever remain static. Thus, the members of a community must make their own decisions about their desired lifestyle and how to achieve it in a sustainable manner because they must live with the consequences of their decisions.

Outsiders can only offer knowledge, expertise, or advice, and if it is accepted, then provide coaching in its use and implementation. It is not uncommon to sit down with community members to test a decision that, although it appears to be 100% correct based on good research and expert advice, proves to be 100% wrong *for now* in terms of the community's holisticgoal. The ability of the holistic decision-making process to help uncover this kind of information makes the holistic process a profoundly useful tool for both development agencies and the communities they are assisting.

Development Principle 4: Handouts Result in a Loss of Dignity

By *handouts* we are not thinking of food and medical supplies in times of disaster, but of goods or services provided in the name of development, which people could have provided or contributed at least partially themselves. When people pay (in money or barter) for something or contribute their labor to some effort, they have ownership of the results, and that is empowering.

When, however, something is simply given to people, the giver retains ownership of that which is given. How many pumps or tractors, in how many communities, lie in rusting heaps because their "owners" failed to maintain them? Far better to allow people the retention of their dignity by owning something they have earned, either through direct payment or labor—and thus understand and truly value. Either way, the people empower themselves through their personal actions to gain control of their lives and destinies.

Strengths and Weaknesses

The greatest strength of the Holistic Management framework is its ability to help people see that they best serve their own interests by accounting for the environmental, as well as the social and economic, consequences of their decisions. The decision-making process incorporated in the framework is simply common sense.

It is not easy to implement "new" ideas, however, even if they are common sense, because there are aspects in the framework that fly in the face of thousands of years of engrained beliefs, such as using livestock to reclaim lands they helped to destroy. That aside, change is seldom easy or smooth for any person, not to mention a community. The changes implied in the holistic framework can also be especially frightening to people who find reactive management more comfortable than assuming the responsibility inherent in proactive management.

This has implications for sustainability and is why short-term development efforts where behavioral change is critical so rarely succeed. Under stress, people will revert back to more comfortable, old behaviors. The case studies discussed below are considered one-hundred-year projects, even though funding has thus far, in the second case, been very short term and disaster related. Even had it been otherwise, it would be hard to call an effort sustainable until the changed behaviors had continued through at least three generations.

Case Studies

Dimbangombe Ranch, Victoria Falls, Zimbabwe

The Africa Centre for Holistic Management owns a property in Zimbabwe. The property, known as Dimbangombe Ranch, includes 2,600 hectares of bush and grassland that is home to a diverse wildlife population, which includes elephant, buffalo, zebra, sable, waterbuck, impala, kudu, lion, leopard, African wild dog, and hyena. The property serves as a training ground for those seeking more sustainable ways of caring for the land, in particular for the people in the neighboring Hwange Communal Lands—145,000 people on over 405,000 hectares.

This case study looks at the goals, strategies, and outcomes related specifically to the Dimbangombe Ranch. Two goals were established:

- To reverse the ecological degradation and enhance biodiversity on Dimbangombe Ranch
- To demonstrate the above in a manner that can be easily replicated, and thus adopted by villagers in the Hwange Communal Lands and elsewhere in Africa

The Strategy

The ranch has a holisticgoal that speaks to the aspirations of the Africa Centre's staff and board of trustees, including a description of a future landscape that is vastly improved over what they started with in 1996. The managers realized that in order to move toward that future landscape and achieve the first goal (ecological restoration), livestock would be needed. The wildlife that remained on the ranch weren't functioning as they once did—in much larger herds continuously on the move due to the presence of large prides of lions. Persecution of the lions, increased settlements, a railway line, as well as roads and other disturbances, had long ago changed these behaviors.

When originally purchased, the ranch was considered fully stocked with one hundred head of cattle. In the best grazing areas, over 75% of the soil surface was bare, and apart from one spring-fed permanent pool, the main river no longer had perennial flow. Between five hundred and one thousand cattle were needed to get a herd big enough to help regenerate the vegetation, soils, rivers, and underground water, as well as the wildlife populations. The Holistic Management decision-testing process helped to find a solution that also met the need of the second goal in making this demonstration relevant: the Hwange villagers were invited to bring their starvation-threatened animals to Dimbangombe to help with this project by joining the herd.

To enhance, rather than disturb, wildlife populations, the decision was taken to forego any fencing and herd the livestock, which included both goats and cattle, to better reflect the situation in the Hwange community. In addition, most of the herders were hired from the Hwange community. To protect livestock from predators, which needed to increase, a lion-proof enclosure (referred to locally as a *kraal*) was built to contain the livestock at night. Made of plentiful local materials—small saplings collected from the bush and woven into portable mats—the kraal design would be readily adopted in the Hwange villages. And because it was portable, the kraal could be moved to where the people wanted their herd to spend the night.

Livestock movements were governed by planned grazing, which becomes even easier to implement when livestock are herded, again better matching the situation in the Hwange community. Photo monitoring points were established, where we hoped to see the most change—areas of bare ground, riparian areas, and in the *vleis* (meadows).

The Outcomes

Over the last eight years, livestock numbers increased almost fourfold on Dimbangombe Ranch, while being herded each day to a grazing plan and confined to the predator-proof kraal at night. (Livestock numbers have fluctuated, depending on the numbers of village cattle brought to graze. The herd has not included less than two hundred cattle in the last six years, with as many as five hundred. Goat numbers in the herd have fluctuated between two hundred and seven hundred.)

The Dimbangombe River now holds water throughout the dry season in all but the driest of years. During the 2004–2005 rains—the driest of the past thirty years—despite no surface flow into the river at all, so much forage grew that the ranch offered to take in an additional one thousand head of cattle from the Hwange community, where animals were again starving. In 2006–2007, during another equally serious meteorological drought, the ranch was able to run five hundred head of cattle and seven hundred goats through to the end of 2007 with no shortage of forage, although surface water was scarce and wildlife, including elephant and buffalo, numerous.

Because of the increase in forage, and grazing planning that has included the requirements of various wildlife populations, there has been a substantial increase in elephant, buffalo, zebra, impala, kudu, reedbuck, bushbuck, and other species. In addition, fish, eagles, ducks, otters, and crocodiles are present for the first time in many years. And because former wetland areas that had dried completely are once more wet and contain large reed beds, it is possible that with continued improvement the Dimbangombe River flow might be maintained even in the driest years.

Pleased by the general improvement, we were still anxious to see what would happen if the treatment with livestock was intensified on a few areas, where ground had remained bare despite the planned grazing. Opinions varied on why nothing had grown on the bare plots. Some thought the past damage by overgrazing and continual trampling rendered them beyond reclamation. Others suggested that the plants surrounding the areas exuded poison from their roots and so prevented anything else from growing. To find the ecological reasons, tests were thus initiated on three, approximately half-acre plots, where the land had been bare since the early 1970s (Figure 7.2).

FIGURE 7.2
Photo taken in September 2004, showing land that had remained bare for over thirty years. Overgrazing of plants had been stopped approximately eight years before this photo was taken, but the ground remained bare regardless of season, each year being a "drought."

FIGURE 7.3
September 2006 view of the same site shown in Figure 7.2.

The lion-proof kraal was moved onto these areas for a week in each case, so that each night the herd of about 350 cattle and 200 goats spent in it, the land was heavily trampled, the soil crust broken, and a lot of dung and urine deposited. The area was then abandoned until the rains commenced, although very little fell during the 2004–2005 rainy season (200 mm of an average 750 mm)—the driest in thirty years. Nevertheless, the treatment worked. For the first time in three decades plants grew once more, confirming our belief that the problem had been too little animal impact. It also confirmed the suspicion that improper management practices produce land deterioration—not droughts.

This area continued to be grazed by the herd and received no special treatment over the next rainy season as part of the routine grazing plan. The rains of 2005–2006 were average, just over 700 mm, but were well distributed throughout the season, and the areas continued to improve (Figure 7.3).

Enhancing Land and Livelihoods in the Hwange Community, Zimbabwe

In 2005, the Africa Centre for Holistic Management received a grant of $426,000 from USAID's Office of Foreign Disaster Assistance, which allowed it to work with the people in the Hwange Communal Lands. Because these were considered disaster relief funds, we were given the limited time frame of one year, later extended to eighteen months, to show results. Data were collected and analyzed at the end of 2006, and a number of modifications to the program made as a result. However, only minimal funds were available to continue the work until a second USAID grant of $329,750 was awarded one year later, in September 2007.

This second grant allowed expansion into two additional communities and the chance to correct many of the mistakes made in the two pilot communities. Because Zimbabwe was suffering yet another meteorological drought, the emphasis in the second grant was on mitigating the effects of the existing drought and the risk of future droughts. The project goals were modified slightly but in essence did not change.

This case study lists the goals for the pilot initiative of 2005–2006, and discusses strategies as well as the outcomes and the modifications they precipitated in the second project phase initiated in 2007. Additional changes made, as a result of the experience gained in 2005–2006, are discussed in "Lessons Learned." The following three goals were identified:

- Use the livestock (cattle, goats, sheep, and donkeys) in two pilot communities to ensure sustainable livelihoods by restoring the productivity of land affected by desertification and shrinking water resources.
- Reduce hunger, while elevating approximately twelve hundred individuals within the most vulnerable families out of poverty through the conversion of an ongoing micro-credit program to one based on livestock currency (due to hyperinflation).
- Effectively address the long-standing cultural prohibition against female ownership of livestock, the issues of safe family sexual practices, and the stigma of those living with HIV through gender empowerment training to community leaders and the 240 male and female household heads participating in the goats-as-currency banks.

Strategy 1: Land Restoration through Grazing

The first step was to identify the communities within the vast communal land area that would serve as the two best pilot communities to employ livestock grazing as a tool for land restoration. The five traditional chiefs who govern the area all serve on the Africa Centre's board of trustees, and their participation and support made this task easier. A survey was conducted to select the two communities based on (1) the number of people already exposed to Holistic Management or who had participated in the Africa Centre's women-only micro-credit program, (2) the presence of well-defined areas with a contiguous boundary used for grazing, and (3) functioning wells (boreholes) that could supply sufficient water to accommodate a large herd.

Four communities were then selected for additional training. Their leaders (formal and informal) toured their own land and were then taken to Dimbangombe to see the differences in the land, talk to the herders, visit the kraals, and see for themselves that combining animals of different species and classes (cows and calves, etc.) into a single herd had no ill effects on individual animals. A series of workshops for livestock owners and community leaders followed this visit to garner a series of commitments: after forming a community goal, they needed to (1) test the decision to run all livestock in a single herd; (2) build a community kraal; (3) initiate a "water fund" with contributions from each family in the community to pay for the maintenance of a pump, reservoir, and troughs; and (4) determine if and how herders would be compensated for their work. The two pilot communities finally selected—Monde and Sianyanga—made these commitments and their leaders, including the chief, signed an agreement and made it public through a ceremony held in each community.

Both communities formed a grazing committee, which met monthly, elected its officers, and selected ten herders. An additional two herd monitors were chosen to evaluate herder and herd performance. Both herders and herd monitors then received training in planned grazing, herding, and animal health at Dimbangombe.

Strategy 2: Goats-as-Currency Banking

The aforementioned commitments were essential perquisites to the goats-as-currency banks. Each of the two pilot communities selected the families who would participate in the goats-as-currency banks. The plan was to select families based solely on need—the most vulnerable in the community. But community members made a strong case for having a mix of vulnerable families (headed by an orphan or grandparent) and entrepreneurial families who could provide coaching to the others. The mix was accepted and became approximately 70% vulnerable families and 30% entrepreneurs, with some of the entrepreneurs acting as guardians for orphans who would become the goat owners.

With inflation in Zimbabwe running officially at 9,200% (much higher unofficially), paper money could no longer serve as currency. Livestock hold their value and produce offspring, so value grows. Goats serve better as bank currency than cattle because turnover is rapid—they can breed at six months and produce two crops of offspring per year. Each bank is composed of twenty families who received half-day training over six consecutive weeks, over which time they created a holisticgoal for their bank and learned to test decisions toward it, such as keeping their goats in the community herd. Further, they (1) formed a banking committee that created bylaws and assumed responsibility for enforcing them, (2) elected bank officers and defined their duties, (3) learned how to manage their banks (through monthly bank meetings), and (4) made commitments on interest rate and repayment of principal, and acceptable uses of funds from the sale of male progeny (to feed their families, pay school fees, or invest in women-owned enterprises). These commitments became part of the bylaws.

In the meantime, the Africa Centre staff set up a bank trust to serve as the "lender," and began purchasing goats for delivery to the banks. Four banks were formed initially—two in each pilot community. The bank trust then contracted with each bank to deliver two hundred female goats (ten for each member family), with the principal to be repaid to the trust at the end of three years. The 30% interest rate was to be paid in goats (two-thirds female, one-third male) each year until the principal was repaid in full, the plan being to use repayment goats to launch additional banks.

Strategy 3: Gender Empowerment

Because the goats-as-currency banks involved whole families, not just women, it became especially important that issues of gender be addressed. Although women could legally own livestock, many weren't aware of their legal rights because livestock were traditionally inherited by men, or boys, rather than women.

Moreover, any discussion on gender also had to address HIV, which is very much an issue of gender empowerment. Many of the village women reported that when they even suggested their partners use condoms, they were beaten. HIV stigmatization is an issue because it prevents people from getting a diagnosis and seeking help, which results in the ostracism of AIDS sufferers and their families, some of whom are bank members. The HIV rate in the two pilot communities was estimated to range from 30 to 50%.

Both a gender specialist and an HIV specialist were retained to train the Africa Centre staff, who in turn provided the training to bank members in one of the five local languages. The training was delivered throughout the general bank training sessions, and reinforcement training continued in the subsequent monthly bank meetings.

Outcomes

In November 2007, when this case study was written, Zimbabwe held a number of unenviable records: the world's fastest shrinking economy, the world's highest inflation rate (officially 9,200%, actually >35,000%), and the world's lowest life expectancy (37 for men, 34 for women). Yet, this project moved forward and did make genuine progress toward the three goals.

Land Restoration through Grazing

The early results were promising in that both communities had combined their animals into single herds by the end of December 2005, and had continued to move them according to the grazing plans they helped create for the wet season (November–March).

This milestone was achieved with some hardship. The villagers had to herd their animals long distances to water over very rough country because watering areas closer to the villages were still under construction. When the herders returned each day, their clothes were in shreds, due to the thorny bushes they'd herded through. Yet they persisted day after day.

However, once the rains were over and crops harvested from fields at the end of April, the commitment to the herding and the grazing plan decreased significantly due to several factors: Neighboring villagers who were short of forage moved their livestock onto the crop fields and grazing areas the pilot communities had reserved for grazing at a later date. Rather than lose all of their crop field stubble to the raiders, many participants withdrew their animals from the community herd to ensure they had access to what remained before it was taken by the raiders. This seriously affected the time calculations in the plan for rationing out the forage over the dry season. In the end, both communities ran short of forage.

The grazing committees reported the infractions to the chiefs, who fined the offenders and in one case instructed a repeat offender to actually leave the area. But enforcement was often too late to save the forage and sometimes arbitrary. This turned out to be a design problem in the project. Enforcement needed to occur at the village level, where it could be dealt with immediately by the village head. This was later rectified as discussed under "Lessons Learned."

Despite the setbacks, there was some land improvement. Areas on which the herds had concentrated were visibly better in terms of forage bulk produced. In general, the pilot communities still had standing forage when surrounding communities had run out. This, as well as the lack of predation losses due to the lion-proof kraals and supervised herding, captured the attention of outsiders. Delegations from six other communities approached the Africa Centre asking if their communities could be part of the project.

This pilot effort also provided an opportunity to demonstrate how livestock could be used to greatly enhance crop yields. As the project was beginning in May 2005, a villager from the Monde community was about to develop a new field for crops and, in the course of a conversation, asked if there was a better way to do it than he was planning—to chop down the trees ("because maize won't grow under them"), bring in oxen to plow with, then plant and apply fertilizer. We said there *was* a better way, and he said he wanted to try it. Knowing that the Monde herd would need to graze his fields as well as others during the dry season, this was an opportunity to demonstrate the benefits of using livestock to till and fertilize worn-out crop fields.

Several villagers agreed to loan him their cattle to put on the new cropping area using a small lion-proof kraal to define the area they would "treat" each night. He moved the kraal around the field until he had treated about half an acre, while the livestock trampled the brush and deposited dung and urine. When the rains were due, he planted his maize, including under the trees he had left standing, as well as beans and melons between the maize plants. Then he waited.

The results surprised even us. In the middle of the rainy season the maize plants were 6 to 8 feet tall in the "hoof-tilled" field, dark green, and growing rapidly (even under the trees). In the conventionally managed neighboring field, on the other hand, the maize plants were sparse, yellowish, and about 3 feet tall. When crops from the two fields were harvested, the difference in yield was remarkable—about fifteen times the yield per acre on the hoof-tilled field compared to the conventional field. The difference in the cob size at harvest is shown in Figure 7.4.

It cannot be said with certainty that previously cropped fields will respond as dramatically as the virgin field did to the animal impact. But at least one other woman used the technique on her field, which she did plow, and the yields were significantly better. One unplanned benefit in her case was the fact that bringing the herd, in this case goats only, to the

FIGURE 7.4
The maize cobs on the right were harvested from the hoof-tilled crop field, and the cobs on the left from the conventionally managed crop field.

field eliminated weeks of labor for her in transporting manure to her field and then spreading it.

Goats-as-Currency Banks

The banks were launched with much fanfare, and monthly bank meetings continued, with training sessions included at each meeting on topics that need reinforcing, particularly on gender and HIV. The plan to reduce hunger, while elevating twelve hundred individuals in the most vulnerable families out of poverty over the food-critical months of October through April, could not be met as hoped during 2005–2006, due to an overambitious project timeline. Nevertheless, eighty families (about five hundred individuals) borrowed ten goats each, which did provide some with greater security over the most food-insecure months in the following year.

The Hwange community members define a poor family as one having fewer than five livestock (of any species), whereas a wealthy family has more than ten animals. Very few families were able to increase numbers through births; many families actually ended the first year of their loan with fewer animals than they had borrowed. This result was again largely a project design problem and also the result of an overambitious timeline. The animals had been purchased in a hurry and delivered to communities without adequate time to adjust to a different environment and a large communal herd, or to be treated for diseases that only manifested after purchase. Training in animal health care had also been insufficient; the herders who received the training were seldom the ones actually herding. And contributions into the animal health care fund were insufficient to purchase medical supplies when needed. Funds collected had not immediately been invested in hard goods, such as livestock, and quickly lost value due to hyperinflation. Finally, there were insufficient herd sires to ensure good breeding coverage until the project loaned each community seventeen male goats. As a result of this series of errors, interest was forgiven for the first year; the lack of production wasn't the fault of the bank members. The interest rate was also reduced from 30% to 10% per annum.

Gender Empowerment

The monthly bank meetings have become one place where both men and women can openly discuss gender issues, HIV prevention and stigmatization, and where women feel confident (and are not punished) for expressing their differing views. An essential step forward for family members (both male *and* female) in the banking program was the acknowledgment that women could own livestock. The loan contracts signed by each bank member ensured that women could also inherit livestock. In fact, some of the bank contract holders were women, even, in some cases, when they had a male partner who was part of the project.

Other signs that appeared to show movement in gender empowerment were: (1) women were selected to serve as bank and grazing committee officers, (2) they were also selected to serve as herders (an occupation traditionally reserved for males), and (3) there was a marked change in women's participation in decision-making processes both in the community and at home. However, there were some reports of women's mistreatment at home once their participation in the community meetings was encouraged and they attempted to translate the behavior to the home setting. As a result, training was added in negotiating skills and the women were provided information on local support resources.

Pre- and post-surveys indicated there was a definite increase in knowledge of HIV, its transmission, and the dangers of stigmatization. Evidence of behavior change to protect oneself and family from the disease was less encouraging, but understandable, given the short time frame. What has been interesting is the eagerness of some of the village men to speak to female project staff members about issues surrounding HIV that they aren't yet comfortable talking about to their spouses.

Lessons Learned

Much has been learned, and much remains to be learned, as this will remain a pilot project through 2009. By then, the two original communities and the two added in 2007 will be much further down the road, and the amount of data will be sufficient enough to draw conclusions that can become guidelines for others seeking to replicate this project. This aside, what have we learned so far?

Involve Community Leaders at the Outset

It is essential that formal leaders (traditional and local government) and informal leaders (opinion formers) be involved at the outset. Once they are committed, implementation will begin to flow. In the Hwange project chiefs were relied on more heavily than lower-level village heads, who were in fact in a better position to enforce agreements and call people together for meetings. In 2007 this was rectified by approaching village heads first and enlisting their assistance in project planning and design. Another error made was in not including local political leaders from the beginning in an attempt to remain politically neutral in a politically divided country. Seeking their support for the next phase of the project in 2007 proved essential to its successful launch. These local leaders are all important in helping to enforce the grazing and banking bylaws, and in championing the case for increased livestock numbers when detractors insist that land improvement can only take place if livestock numbers are reduced.

Engage Stakeholders

It takes an extraordinary effort to fully engage all the stakeholders. Despite many repeat meetings and training sessions to cater for people who were absent in the project's first phase, some livestock owners missed them all, and these people were the first to take their animals out of the herd in the dry season. Those who kept their animals together throughout the entire project had received the most training. This included the villagers participating in the goat banks who continue to meet monthly and receive reinforcement training. To enhance the likelihood of engaging all livestock owners in 2007, training cascades were set up in each community. Project staff serve as master trainers for village-based facilitators (two per twenty families) selected with the help of village heads. Each village-based facilitator then provides continuous training and coaching of the families he or she is responsible for, again with the support of the village head.

The Holisticgoal Is Pivotal in Maintaining Commitment

Commitment must be reinforced through continual reminders. Although the members of each community have recorded their holisticgoal in writing (in local languages), many members are illiterate and verbal repetition is essential. More important than mere repetition is to use the holisticgoal as the reference point for testing decisions, as well as continually monitoring and noting progress toward its achievement.

Self-Governance Has to Be Learned

It isn't safe to assume that communities, or even community leaders, are able to create, manage, and lead committees or other efforts without some guidance or coaching. This was a hard-learned lesson in the early phase of this project when staff were eager to foster the belief in the two pilot communities that this project was "theirs." Toward that end they had grazing and bank committees create their own bylaws, which then had to be revised month after month as omissions were uncovered.

On the other hand, errors were made in the other direction as well. Project staff didn't realize why some village heads in the first two communities had withheld their support. Only at the end of the first year, when they were evaluating the project, did they discover that village heads had felt slighted by the fact they had had little say in program decision making. Villagers had only so much say, especially in selecting the grazing committee or bank members and in forming their own bylaws.

This oversight had made it even more difficult to enforce bylaws. It was corrected in 2007.

Agreements Up Front Are Essential

These agreements should stipulate what the prerequisites are prior to community selection, what each community then commits to doing, followed by the consequences for nonperformance. It is important that the formal leaders in each community understand these agreements well enough to explain them to their constituents and continually reinforce them as they monitor compliance. Project staff must also be willing to enforce the penalties for noncompliance, but only if they are certain the noncompliance isn't a result of their own inadequate performance or failure to honor commitments they have made. The latter point is essential in building trust with the community. For example, when bank goats failed to thrive due to a variety of reasons staff could have prevented had they known better, they didn't hold bank members to their agreement to make their first interest payment. Likewise, when each community agreed to pay into a water-point maintenance fund (to maintain equipment) but failed to invest the funds in a nondepreciating asset, it was more likely due to faulty project design than to community commitment. In 2007 the fund became something only livestock owners contributed to, and the more livestock owned, the greater the contribution because the greater the benefit from the water development and livestock medications. What's more, no money changed hands, only livestock were "pledged" and then the debt called in only when supplies were needed.

The Status of Herders Needs to Be Elevated

The Hwange community is not alone in relegating herders to lowly status. Many other pastoral communities do so as well, giving the job of herding to small children. Yet, when livestock are used as land restoration tools, the role of the herder becomes one of the most valuable there is in a community. The first herders selected and trained for the project enjoyed the training, but because both communities elected not to pay herders and instead have livestock owners take turns herding, the training did little good. In many cases herding duties were turned over to youngsters. The herd monitors who trained alongside the original herders took their jobs a little more seriously, but lack of leadership from the grazing committees they reported to resulted in mixed performance.

As a result of these lessons, a number of changes were made in 2007. Herders were selected with the help of the village heads, and those selected were trained to become senior herders who would put together and train herding teams and report to their respective grazing committee. Senior herders would assist their grazing committee in creating the community grazing plan and would also receive specialized training as community animal health workers. They would be given identifying caps and t-shirts to wear when leading their herding teams, and the grazing committees they reported to would honor the best herding teams with special awards each year. Grazing committees were asked to consider providing some remuneration for senior herders from within the community water and animal health maintenance fund.

Some Community Members Will Feel a Loss of Power

As a result, they may become obstructive when they see the greater community pulling together and working toward a common holisticgoal. Initially, these people were ignored and left behind. Going forward in 2007, they were acknowledged. With the help of the village heads they were given parameters within which they could graze their animals. These parameters included a restricted grazing area around the homestead and a special corridor to water. The plan is to encourage them to join in the project rather than feel unwelcome.

Model Behavioral Changes

It is essential that project staff model the behavioral changes they are seeking. For example, the Africa Centre project leader is a woman who (1) owns livestock herself; (2) can say she and her husband will not be demanding a bride price for their daughter—they will not "sell" her to her prospective husband; (3) can speak to men as equals—chastise them, laugh with them, and above all be firm with them; and (4) can model her own mutually respectful relationship with her husband.

Fluency in Articulating the Reasons for Change Takes Time

It will likely be many years before every member of a community can explain what they are doing and why they are doing it in a way that others can understand. The formal and informal leaders will usually be the first to acquire this ability. In this project it is anticipated that both village-based facilitators and senior herders will either precede or be close behind them.

It is unrealistic to talk about sustaining the outcomes of this project until the change has been carried through at least two, and possibly three, generations. We at the Savory Institute and the Africa Centre have been too frequently caught believing we had sustainable results only to see a property change hands and management regress to decisions based on objectives while the land once more degrades. Ultimately, intergenerational transfer is essential to sustaining any major change in a community.

Conclusion

To be sustainable, human activities must, of necessity, be linked to the maintenance of productive land and the restoration of land so degraded that it can no longer produce the ecological services required by the people who depend on it for their livelihoods. When land is bare and eroding, most of the rain runs off or evaporates, which in turn leads to perennial rivers becoming intermittent and wells (boreholes) and dams running dry. When a community's water resources are lost, so is the community's ability to feed itself or nurture entrepreneurial efforts. Hence, the community is doomed. The Hwange community is well on its way down this path, and it is only one of many thousands moving in the same direction.

On two-thirds of Earth's land surface, where humidity and rainfall are seasonal, grazing and animal impact are the best tools, in fact the only tools, that can be used to begin to restore land and water resources on the vast scale required. In these cases, virtually all the technology developed to do what livestock do for free is fossil fuel dependent. Livestock movements must be planned so that the land is sufficiently impacted but plants aren't overgrazed and the inherent, biophysical complexity of the ecosystem is honored and cared for.

What is heartening is that using livestock as a tool for restoring deteriorating land rarely calls for a decrease in livestock numbers. The problem in most seasonally humid environments is too few animals (livestock or wildlife) on the land. As a result, much of the soil is bare and covered in a crust that decreases aeration and water penetration. The hooves of animals in a concentrated herd will break the crust and encourage plant growth, as long as the animals don't remain in an area too long or return too soon.

There is no need to destroy the pastoralist societies that have long existed in seasonally humid environments, such as the Masai or Samburu in Kenya, by forcing them to leave their ancestral grazing areas and settle on small plots or in towns. Nor is it necessary to reduce the numbers of livestock in an effort to save the land. The political, economic, and development ramifications are staggering given the opportunities it presents.

The education and empowerment of women is critical to any effort to sustainably improve the quality of a people's lives. Although there are advantages in providing training to women only, mainly because they tend to be more forthcoming in women-only groups, men require the same training if women are to attain any measure of social equality. Further, both genders require a context within which they can practice changed behaviors, such as the goats-as-currency banks provided in the Hwange pilot communities. Education for women should ideally start when they are children. One of the most successful ways to create opportunities for the education of young girls is micro-credit programs, such as goats-as-currency banking, that involve agreements as to where profits will be invested—school fees being one of the priorities.

Efforts to achieve sustainability that are framed within the context of goals and objectives only, but without reference to the whole to be managed or a holisticgoal that speaks to both immediate and longer-term needs and desires, are unlikely to succeed. Hence, the Holistic Management approach involves (1) defining the whole to be managed; (2) setting a holisticgoal that describes the quality of life the stakeholders collectively seek, what they must produce to achieve it, and the resource base they depend on *as it will have to be* to sustain their production and quality of life; (3) determining the tools, resources, and knowledge needed to achieve the many objectives and goals in everyday life; (4) testing all potential objectives and actions to make sure that they are socially, environmentally, and economically sound and in line with the holisticgoal; and (5) monitoring the results of decisions made and actions taken to ensure required results and target correction where necessary (Savory and Butterfield 1999).

That said, there is no one way, no one answer, to any social–environmental challenge. Rather, there are millions of answers and potential solutions, and these have to be worked out case by case, situation by situation, by people who are driven by a desire for something better. United behind a holisticgoal they have formed themselves, and armed with the ability to test their decisions toward it, a people can better advise an outside team of experts on what is right *for the people*. Then the people can begin to realize and use the vast knowledge they already have to incorporate the experience and counsel the outside team provides. This approach goes a long way to ensuring that efforts toward sustainable development will be successful—and thus passed from one generation to the next.

Endnote

1. Holistic Management® is a registered trademark of Holistic Management International.

References

Gordon, I., and P. Duncan. 1988. Pastures new for conservation. *New Scientist* 1604:54–59.

Hobbs, R.J., and L.F. Huenneke. 1992. Disturbance, diversity, and invasion: Implications for conservation. *Conservation Biology* 6:324–37.

McNaughton, S.J., and F.F. Banyikwa. 1997. Promotion of the cycling of diet-enhancing nutrients by African grazers. *Science* 278:1798.

Pickett, S.T.A., and P.S. White. 1985. *The ecology of natural disturbance and patch dynamics*. Orlando, FL: Academic Press.

Savory, A., and J. Butterfield. 1999. *Holistic management: A new framework for decision making*. 2nd ed. Washington, DC: Island Press.

Voisin, A. 1960. *Better grassland sward: Ecology, botany, management*. London: Crosby Lockwood & Son Ltd.

Voisin, A. 1962. *Rational grazing: The meeting of cow and grass: A manual of grass productivity*. London: Crosby Lockwood & Son Ltd.

Voisin, A. 1988. *Grass productivity*. Washington, DC: Island Press.

8

Taking Personal Action toward Sustainability

Viviane Simon-Brown and Chris Maser

CONTENTS

Introduction

Sustainable living is an ethic of stewardship in which our desire for fulfilling and productive lives is thoughtfully and consciously balanced with the social, economic, and environmental security of life on Earth, now and for future generations. In other words, it is "living a life that is deeply satisfying, fulfilling, and appealing—and at the same time, environmentally responsible" (Simon-Brown 2008, 11).

We could use the term *sustainable development* to describe sustainable living; however, many people assume sustainable development involves only large arenas of public policy and community development. *Voluntary simplicity, conscious simplicity, intelligent consumption,* and *integrated living design* are other descriptors. People say, "I'm being frugal" or "I'm downsizing," or simply, "I'm simplifying my life." Whatever words are used, over 60% of Americans currently embrace some version of sustainable living (Schor 1998).

The preceding chapters in this volume focused on a variety of ways people can work together to support and promote sustainable development.

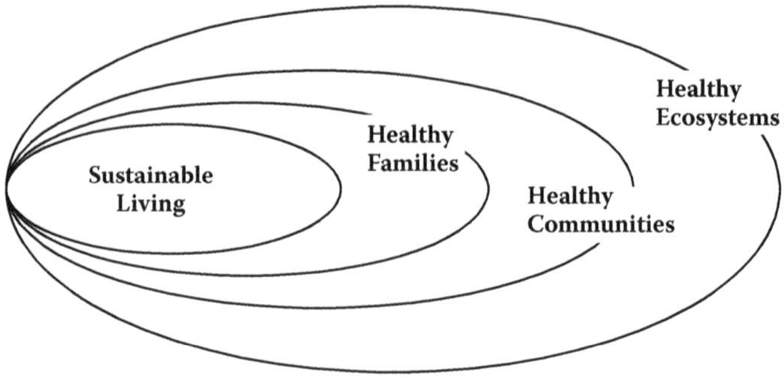

FIGURE 8.1
Sustainable living starts at home and expands outward to families, communities, and ecosystems.

This chapter on sustainable living focuses on *managing the only thing we can manage, ourselves.* Sustainable living creates an essential ethical core to support the concepts of using our natural resources sustainably. Sustainable living starts with an individual and expands outward to families, communities, and ecosystems. See Figure 8.1.

Sustainable living is for mainstream adults, typical Americans with two cars in their garage, one of which might be a sport utility vehicle. It is for people who ski, watch TV, hike, read, snowmobile, go to movies, rollerblade, play video games, bowl, and camp. It is for people with mortgages and those who drop their kids off at child care. And it is also for people with credit cards, who shop at malls, and who eat fast foods. Clearly, anyone can practice sustainable living and would be wise to do so, not only from a spiritual or ethical point of view, but also as an unconditional gift from one generation to all those who will follow.

Sustainable living is about making thoughtful decisions, considering the social–environmental impacts of our consumer choices, and finding practical alternatives to environmentally inappropriate practices and their products. Sustainable living is noncompetitive yet is based on individual definitions of an acceptable quality of life, and thus is deeply personal because individuals have different requirements and values. Sustainable living focuses on what we do today and how our actions will affect tomorrow and beyond. It combines practical ideas, such as turning off lights, with such intangible values as one's personal quality of life. Finally, sustainable living puts our individual actions into a global context because the positive steps taken at the individual and family level affect the world as a whole; put differently, we act locally and affect the whole world. Wackernagel and Rees, in *Our Ecological Footprint,* observed as follows. "It would require four Earths for everybody on the planet to live the lifestyle of North Americans" (Wackernagel and Rees 1996, 160). Why are we living so unsustainably you might ask? To find the answer, one needs to look at the factors limiting the adoption of a sustainable lifestyle.

Barriers to Sustainability

In the United States, for instance, there are three major barriers to the adoption of a sustainable lifestyle: (1) a time crisis, (2) a disconnection with our natural world, and (3) materialism. In addition, there are three learning obstacles to surmount: (4) negative framing, (5) symptomatic viewpoints, and (6) the lack of a common frame of reference. These barriers and obstacles underscore the fact that (7) American society does not embrace a conservation ethic.

The Time Crisis

Time is the most precious commodity in the United States. Consider the following statistics. American workers have less free time than any other industrial nation. Employed Americans spent 142 hours more per year on the job in 1994 than they did in 1973 (Mishel et al. 1997). Americans average an hour and six minutes less sleep each night during the week than the recommended eight hours, according to the U.S. National Sleep Foundation, and almost half an hour less each night during the weekend. By year's end, Americans have lost 330 hours of rest (National Sleep Foundation 2008). Most Americans wish they had more time to spend with family. This sentiment is shared among parents (88%) and nonparents (83%) (Widmeyer Research & Polling 2003). For three- to twelve-year-olds, time spent just sitting and talking with someone at home has been reduced from fifty-three minutes in 1981 to thirty-five minutes in 1997 (Owens and Hofferth 2001).

Time is the great equalizer. We all get the same number of minutes in a day. It is how we choose to spend that time that makes the difference. For example, our time crisis makes it easy to understand why we choose minimal contact with the natural world.

The Disconnection from the Natural World

American adults spend 87% of the day indoors (Klepeis et al. 2000). Our self-image is that of rugged, outdoor individuals, but our nation has become an inside society. We watch nature shows on the Discovery Channel rather than exploring a nearby forest, beach, or desert. We use climbing walls in gyms rather than climbing real rocks. We walk on treadmills rather than along a river. We buy vehicles capable of transporting us into the backcountry, but we go to the mall or the movies instead. In fact, the most visited location in America isn't Yellowstone National Park, with 3.1 million visitors in 2008, or the Grand Canyon, with 4.4 million visitors in 2008 (National Park Service 2008). It's Bloomington, Minnesota's Mall of America, which attracts over 40 million visitors per year (Tvrdik et al. 2006). A significant percentage of American kids rarely come in contact with nature. This is due to electronic devices and games, coupled with a dramatic increase in parental concerns about safety. Richard Louv, in his book *Last Child in the Woods: Saving Our Children from Nature-Deficit Disorder,* introduces a growing body of scientific research that suggests attention-deficit disorder (ADD) and other childhood maladies are the results of alienation from nature. He states that children who are given early and ongoing positive exposure to nature thrive in intellectual, spiritual, and physical ways that their "shut-in" peers do not (Louv 2005, 334).

Materialism

Materialism is another barrier to sustainable living. Materialism "connotes an unbounded desire to acquire, followed by a throwaway mentality," states Juliet Schor, author of *The Overspent American: Why We Want What We Don't Need* (Schor 1998, 217). Americans are confused. We belong to the world's most materialistic society, yet polls indicate our deepest aspirations are nonmaterial (Harwood Group 1995). Most materialism is a vain attempt to fill our spiritual emptiness, which is why we *want the wanting* but soon discard what we have recently acquired.

Per capita income has risen 62% in the United States since 1970 (U.S. Census Bureau 1996). However, while disposable personal income increased, the quality of life decreased 51% in the same time period, as measured by the Index of Social Health (Miringoff et al. 1999).

American consumption far exceeds that of any other country. In fact, the amount of energy used by one American is equivalent to that used by 3 Germans, 6 Mexicans, 14 Chinese, 38 Indians, or 168 Bangladeshi (U.S. Census Bureau 2005). We have less than 5% of the world's population, yet we use 25 to 35% of the world's resources and produce 25% of the world's waste (Flint and Houser 2001). Human consumption consists of more than just food; we require manufactured, nonfood items, such as energy, housing, clothing, vehicles, and a vast array of other goods as well.

Seventy-five percent of Americans have credit cards, with 46% of these carrying a median balance of $2,200 each month, according to the 2004 Federal Reserve Survey (Bucks et al. 2006). Ironically, the industry jargon for someone who pays credit card bills in full each month is "deadbeat." Sixty percent of American families have so little savings that, if they lost their jobs, they could not sustain their lifestyle for more than one month.

And, about 43% of American families spend more than they earn each year.

Negative Framing

With respect to living sustainably, there is great power in learning to reframe negatives into positives. Much of our daily confusion in communication comes from trying to move away from negatives. As long as people express what they do not want, it is virtually impossible for them to figure out what they *do* want. (Maser 1998).

Symptomatic Viewpoints

Clearly, how we think determines what we do. A narrow, economic focus on individual pieces of a system is like a racehorse with blinders that prevent it from seeing anything but the racetrack in front of it. A linear-minded problem-solving approach to a dynamic, interactive, cyclical system leads to a symptomatic rather than a systemic view.

In contrast, a systems thinker (a person who sees the whole in each piece) is concerned about tinkering willy-nilly with the pieces because they know such tinkering might inadvertently upset the desirable function of the entire system. Benjamin Franklin, in his *Poor Richard's Almanac*, expressed it simply: "For want of a nail, the shoe was lost. For want of a shoe, the horse was lost. For want of a horse, the rider was lost. For want of a rider, the battle was lost" (Maser 2005).

The Lack of a Common Frame of Reference

The lack of a common experience or frame of reference is probably the greatest obstacle to effective communication. Although many people believe that words carry meaning in much the same way as a person transports an armful of wood or a pail of water from one place to another, words *never* carry precisely the same meaning from one person to another. Words are vehicles of perceptive meaning and may or may not supply emotional meaning as well. The nature of a person's response in any type of communication is determined by his or her past experiences surrounding the word and the feelings it evokes (Maser 1996).

The Lack of a Conservation Ethic

Americans do not have a unifying conservation ethic. Only two slogans are universally known: Woodsy Owl's "Give a hoot, Don't pollute!" and the original Smokey Bear mantra, "Only YOU can prevent forest fires." In fact, the "Report of the Intelligent Consumption Project" indicates that developing a viable conservation ethic in the United States is central to shaping our conservation policy and practice (Strigel and Meine 2001, 39). So, the question is: How do we learn to live sustainably?

The Primary Actions Required for Sustainable Living

Personal values strongly influence decisions; therefore, we have to first acknowledge the role values, ethics, and beliefs play in our lives. Dr. Courtland Smith, emeritus professor of anthropology at Oregon State University, provides the following distinctions between values, ethics, and beliefs (Smith and Gilden 2000).

Values are the lens through which we perceive reality. When we are faced with a choice, that lens shows us doing right or wrong, being desirable or undesirable, being good or bad, taking appropriate or inappropriate actions, and so on—all based on the predetermined roles assigned by the culture we live in. Most of our personal values, which we identify by the time we are twelve years old, are derived largely from interactions with family, friends, teachers, and community members. Values can change over time, however, in such places as high school and college, or in unusual situations, such as traveling in other countries. Almost all of our day-to-day decisions are based on our value system, and thus have implications for our individual concepts of sustainable living.

Ethics are value sets. Certain values are usually found together, as exemplified by cars in a parking lot. Certain makes of cars will generally have readily recognized bumper stickers. Volvo station wagon: "baby on board," "Obama/Biden," "my child is an honor student at…." A full-size Ford pickup with dual tires, on the other hand, would most likely have bumper stickers such as "NRA," "cowboy and proud of it," "my border collie is smarter than your honor student." While these are generalizations, consider this: Have you ever seen a "save the whales" bumper sticker on a Humvee?

Facts are actually beliefs, what we perceive to be true. There are two parts to the definition: facts are (1) provable, measurable, and (2) socially agreed upon. Fact: The world is flat. This was absolutely true in the eleventh century. You could prove it by climbing the nearest church tower, and everyone knew it. Beliefs can change over time, but the transition is messy and painful, since we, consciously or unconsciously, choose our facts to meet our perceived needs.

Our values, ethics, and beliefs are all encompassed in our decision making. In other words, our decision making is based on our values, ethics, and beliefs.

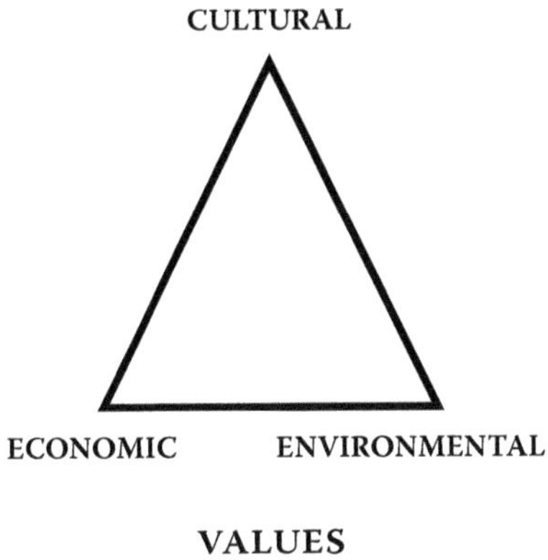

FIGURE 8.2
The sustainability triangle. (From Simon-Brown, V. 2004. *Journal of Extension*, 42(5). With permission.)

The first step would be to identify our individual core values by using a *sustainability triangle* (see Figure 8.2). A sustainability triangle typically identifies the cultural, economic, and environmental aspects of an issue. In business, this is called the triple bottom line, an increasingly popular notion of three fully integrated goals: (1) increase profits, (2) enrich the planet, and (3) enhance the lives of people.

It's important to note that a sustainability triangle is equilateral. Balancing our economic, cultural, and environmental values is the whole idea of sustainable living.

It is helpful to identify three to five top personal values for each side of the sustainability triangle, cultural, economic, and social, in order to live a life that is deeply satisfying, fulfilling, and appealing—while at the same time being environmentally responsible. Once selected, these values are employed in all decision making if true sustainability is to be a viable option. For example, cultural values statements might include: my friends and family are of the utmost importance; all people are treated equally and respectfully; life is enhanced by diversity; I am an integral part of my community of place. Economic value statements might include: I avoid impulse shopping; I live within my means; I support my community by purchasing locally as much as possible; I know how much is enough. Environmental value statements might include: I recognize that the natural environment has intrinsic value; I work for long-term sustainability of natural resources; I am accountable for my actions. Choice is the most important thing we have to bequeath to future generations, and I am doing my part to ensure there are values and choices to pass forward. A central concept for sustainable living is that individuals have to decide not only what is important to them, but also what makes sense to them.

Taking the time to consciously consider our strongest values, our deepest beliefs, and the consequences of our actions will result in more balanced and reasoned decisions, as well as more sustainable outcomes. This is, at the individual level, a conservation ethic in the making.

References

Bucks, B.K., A.B. Kennickell, and K.B. Moore. 2006. Recent changes in US family finances: Evidence from the 2001 and 2004 Survey of Consumer Finances. http://www.federalreserve.gov/pubs/oss/oss2/2004/bull0206.pdf (accessed April 12, 2008).

Flint, R.W., and W.L. Houser. 2001. *Living a sustainable lifestyle for our children's children*. Lincoln, NE: Authors Choice Press.

Harwood Group. 1995. Yearning for balance views on consumption introduction. http://www.iisd.ca/consume/harwood.html (accessed April 12, 2008).

Klepeis, N.E., W. Nelson, W. Ott, J. Robinson, A. Tsang, P. Switzer, J. Behar, S. Hern, and W. Engelmann. 2000. *The National Human Activity Pattern Survey (NHAPS): A resource for assessing exposure to environmental pollutants*. Berkeley, CA: Ernest Orlando Lawrence Berkeley National Laboratory.

Louv, R. 2005. *Last child in the woods: Saving our children from nature-deficit disorder*. 1st ed. Chapel Hill, NC: Algonquin Books of Chapel Hill.

Lovins, A. 2002. Natural economy. *Resurgence* 213:16–18.

Maser, C. 1996. *Resolving environmental conflict: Towards sustainable community development*. Delray Beach, FL: St. Lucie Press.

Maser, C. 1998. *Vision and leadership in sustainable development*. Boca Raton, FL: Lewis Publishers.

Maser, C. 2005. *Our forest legacy: Today's decisions, tomorrow's consequences*. Washington, DC: Maisonneuve Press.

Miringoff, M.L., M.-L. Miringoff, and S. Opdycke. 1999. *The social health of the nation: How America is really doing*. New York: Oxford University Press.

Mishel, L.R., J. Bernstein, and J. Schmitt. 1997. *The state of working America, 1996–97*. Washington, DC: Economic Policy Institute.

National Park Service. 2008. http://www.nature.nps.gov/stats/index.cfm.

National Sleep Foundation. 2008. Sleep in America poll, summary of findings, a national survey of adult Americans who work at least 30 hours per week for pay. http://www.sleepfoundation.org/atf/cf/%7Bf6bf2668-a1b4-4fe8-8d1a-a5d39340d9cb%7D/2008%20POLL%20SOF.PDF (accessed April 12, 2008).

Owens, T.J., and S.L. Hofferth. 2001. *Children at the millennium: Where have we come from, where are we going? Advances in life course research*. Vol. 6. Amsterdam: Jai.

Schor, J. 1998. *The overspent American: Why we want what we don't need*. New York: Basic Books.

Schor, J., and B.S. Taylor. 2002. Cleaning the closet. In *Sustainable planet: Solutions for the twenty-first century*, ed. J.B. Schor and B. Taylor. Boston: Beacon Press.

Simon-Brown, V. 2004. Intelligent consumption: Addressing consumer responsibilities for natural resources—and beyond. *Journal of Extension* 42(5). http://www.joe.org/2004-october/a1.shtml

Simon-Brown, V. 2008. Sustainable Living Project, Forestry Extension Program, Oregon State University. http://www.cof.orst.edu/cof/extended/sustain/ (accessed April 11, 2008).

Smith, C.L., and J.D. Gilden. 2000. *Values: The lens through which we view reality.* Corvallis: Oregon State University, Department of Anthropology.

Strigel, M., and C. Meine. 2001. Report of the Intelligent Consumption Project: A collaborative project of the Wisconsin Academy of Sciences, Arts and Letters and the USDA Forest Service. http://www.fpl.fs.fed.us/tmu/wood_for_energy/intelligentconsumptionproject.pdf (accessed April 12, 2008).

Tvrdik, M., K. Robb, and N. Medvec. 2006. Mall of America marketing opportunities. http://www.mallofamerica.com/DAM_public/4243.pdf (accessed April 12, 2008).

U.S. Census Bureau. 1996. *Statistical abstract of the United St*ates. http://www.census.gov/prod/2/gen/96statab/96statab.html (accessed April 12, 2008).

U.S. Census Bureau. 2005. *Statistical abstract of the United States 1995–2000.* http://www.census.gov/prod/www/statistical-abstract-1995_2000.html (accessed April 12, 2008).

Wackernagel, M., and W.E. Rees. 1996. *Our ecological footprint: Reducing human impact on the Earth.* New Catalyst Bioregional Series, Vol. 9. Philadelphia: New Society Publishers.

Widmeyer Research & Polling. 2003. Take back your time poll. http://www.newdream.org/about/polls/timepoll.php (accessed May 25, 2009). Based on a nationally representative telephone study of five hundred American adults. Margin of error ±4.4%.

9

Participatory Leadership for Sustainable Development[1]

Okechukwu Ukaga, Charles Moumouni, Michael Reichenbach, and Chris Maser

CONTENTS

The inspiration for this book came from a capstone class that Okey Ukaga designed and taught at the University of Minnesota, Duluth. The course, Participatory Process for Sustainable Development (PPSD), provided enrollees the opportunity to understand, evaluate, and apply basic principles of sustainable development and participatory project planning.

One aspect of the class was to help students develop and demonstrate an understanding of sustainable development principles and frameworks. Another aspect involved helping students explore the application of participatory processes. Against this background, Ukaga wanted to create a book that highlights theoretical approaches to and salient examples of sustainable development. Hence, he invited two professional colleagues, Chris Maser and Michael Reichenbach, to coedit this volume with him.

We organized a team with both theoretical and practical experience in selected aspects of sustainable development to contribute to this edited volume. The previous chapters relate to sustainable development principles and frameworks. In this chapter, we focus on the role of participatory leadership and communication in sustainable development.

Sustainable Development and Participatory Leadership

While there are many opinions of what defines sustainable development, it is generally agreed that sustainable development is oriented around the future as well as community. The World Commission on Economic Development defined sustainable development as seeking to meet today's needs without compromising the ability of present or future generations to meet future needs (WCED 1987). Working in sustainable development may mean putting aside personal desires and working toward the good of the community. To accomplish this, participatory project planning and implementation is imperative. However, it often comes against the hard reality of a culture where top-down programming is the norm. The challenge is to recognize this problem of nonparticipatory tendencies and work to mitigate it so that active participation and effective partnership of all stakeholders can take place, and in turn lead to sustainable development.

Servant Leaders and Active Citizens

True participation is based on joint ownership of both the process and the outcome by each and every participant. This enhances intimate personal experience as each stakeholder involved commits to both the process and the outcome. For the past thirty years, development scholars and practitioners have been focusing their attention more on what a group or community of people can do as whole. While group-oriented actions should remain the paramount goal, the potential of the individual to lead and bring about change should not be neglected (Figueroa et al. 2002). To achieve true participation that can lead to sustainable development, therefore, we need "servant leaders" and "active citizens" who can create and maintain safe emotional and political environments within which people can develop good relationships and work together to plan and implement mutually beneficial projects.

Service to the community—not just information and persuasion—is the true mark of servant leaders. Such leaders, as the late Robert K. Greenleaf rightly noted, want to make sure that the people's needs are met; that those being served grow and become wiser, better, more self-reliant, and more likely to become servant leaders themselves. Further, servant leaders operate out of a moral principle of justice that benefits the least placed in the society (Greenleaf 1977).

Thus, unlike opinion leaders, the main role of servant leaders is not to simply diffuse innovations or transfer technologies, but to enable people to envision a better future and work together to achieve it (Lazarsfeld 1944; Rogers 1995). The servant leaders belong to the community, and have the ability to tap into the potential of every one of their people, catalyzing their personal strengths and potentials. More importantly, servant leaders are motivated by their desire to ensure that the people's most important needs are met, and that those they are leading grow and are able to meet their own needs.

Both servant leadership and active citizenship are essential components of the participatory leadership model. Active citizenship calls for all members of the community to be actively involved in all aspects and stages of the design and implementation of policies and projects that affect them. Servant leaders and active citizens act on what they believe. Thus, they differ from other people who may have goodwill but fail to act to make a positive difference. "Where both active citizenship and servant leadership are present, sustainable development is possible due to participatory program planning, implementation, monitoring and evaluation based on genuine concern for and involvement of the people" (Ukaga 2005, 239). Hence, we envision a participatory development communication approach that makes tandem and appropriate use of servant leadership and active citizenship.

Participatory Leadership Communication for Sustainable Development

The Participatory Leadership Communication for Sustainable Development (PLCSD) approach is an iterative process by which servant leadership and active citizenship work together to promote sustainable development by taking a good idea from one or more person(s) and moving it forward through a growing "spiral[2] of communication" that eventually involves all stakeholders in a collective action to meet an identified need.

Step 1: Conceptualization. The approach is based on the simple assumption that an idea for any good effort must come from somewhere, and typically such an idea would come not from all stakeholders at the same time, but instead from one or a few individuals or stakeholders initially. Anyone can have a great idea and take initiative.

Step 2: Validation. For the person with the idea to do a preliminary reality check in terms of appropriateness, efficacy, and technical feasibility by sharing the idea with a few people within the group and possibly with outside experts. This is done to get early feedback regarding the idea. Key channels of communication at step 2 include interpersonal (one-on-one) tools as well as congresses, seminars, workshops, and focus groups.

Step 3: Legitimating. Assuming that the idea is still deemed viable and appropriate, legitimating would involve sharing and double-checking the idea (or an improved version due to input from the previous step) with the whole community and evaluating both the need for the idea and its appropriateness. Depending on the type of project, the size and nature of the community, and the spatial and temporal scales involved, communication methods used at this stage can range from simpler methods, such as meetings and focus groups, to more complex or comprehensive efforts, such as participatory appraisal techniques (Ukaga and Maser 2004). Whereas mass media are the key channels to create popular awareness of the idea, indigenous communication tools (theatre, song, dance, storytelling, etc.) should be preferred to television, national radio, and newspapers in some contexts. Alternative media like video, community radio, and newspapers in local language should also be explored. This is the stage at which the community should be involved to ensure that the community understands the idea very well and makes an informed decision as a collective regarding if and how to proceed with the implementation of the proposed idea or initiative. It is pertinent to note that not all ideas make it through all the stages. An idea may be dropped at any stage for lack of support or the realization that it is not appropriate or feasible. However, when done properly, an idea gathers more momentum and wider support and becomes more likely to be implemented successfully as it moves from step 1 to step 3 and beyond. It should be understood that step 3 is the heart of the participatory leadership communication process.

Step 4: Implementation. Implementation involves a noticeable transformation whereby the community not only buys into the idea, but also assume co-leadership through collective action in implementing the initiative. This step can be visualized through the metaphor of the flying geese. As the idea gains more support and reaches the point of popular involvement, people are ready not only to follow the leader, but to lead as well. Throughout this iterative process, and especially from steps 3 to 5, stakeholders can step in and out of both servant leader and active citizen roles as necessary and appropriate to move the process forward and get the job done.

Step 5: Appropriation. Appropriation involves owning, maintaining, and sustaining the collective action and ensuring that the initiative is successful and effective. This should include formal evaluation of not just the inputs, outputs, outcomes, and impacts, but also the processes and the various roles played by the servant leaders and the active citizens. This will help document what works, how and why, and what does not work and why, which will in turn lead to learning and continuous improvement. At this stage, the idea is clearly no longer that of one or a few persons from whom it originated, but has been transformed into the group's idea and most likely was revised and changed considerably through the iterative communication process or community dialogue to reflect the thinking and wishes of all stakeholders.

Participatory Leadership Communication for Sustainable Development Examples

The following examples demonstrate the importance of the PLCSD approach.

Chicoutii, Canada

Many people around the world might not have noticed what happened at Chicoutimi in the region of Saguenay, Canada, after the devastating flood there in 1996. After that flood, the government, disregarding the opinion of the affected people and their feelings, appointed experts to study the event and recommend viable solutions. The experts recommended that the flood-risk area be turned into a national park. Government, on the basis of that recommendation, decided to relocate the people to a "better" area with "better" housing and amenities.

This was done without any consultation with the people to determine what they really wanted. The result is that up till today, many of the people feel rather uprooted. The implication of this is that, laudable as the idea of moving the people to a "safe" location might seem, it may not be sustainable in the long term. This is because many of the people feel that they have lost their identity by the forced relocation. The situation may lead to some of them eventually abandoning the new location. If the PLCSD model were applied, the result would certainly have been more positive, as the next example shows.

Valmeyer, United States

The Chicoutimi event contrasts very sharply with what happened in Valmeyer, Missouri. Unlike the inhabitants of Chicoutimi, those of Valmeyer voluntarily relocated to a new settlement after their village was completely submerged in water in 1993 following a heavy flood. The mayor of Valmeyer got the idea that the whole community could relocate to an area that is relatively free from the risk of being submerged in floodwater. He began talking to the people about it, and then the county's regional planning committee drew up a number of options for the future, one of which was total relocation of the community as one. The whole community chose to relocate. In the new location, they each chose where to build their own homes, and together, through consensus, where to build the public facilities like the markets, the churches, the parks, etc. The naming of the streets and facilities was done also through a participatory process.

The crucial difference between Chicoutimi and Valmeyer is that in Valmeyer, the idea of leaving—though one person's idea in the first instance—was successfully communicated to the whole people, who validated and effectively implemented it. As a result, the inhabitants of Valmeyer are very happy being where they are because it is "their" idea and they have appropriated it. The feeling of satisfaction that now prevails is because of the mayor and his city council's participatory leadership. Valmeyer is a good example of what happens where the five steps of the PLCSD approach are implemented.

Radio Sutatenza, Columbia

The example of the first well-known community radio project, Radio Sutatenza, established in 1947 and which subsequently influenced participatory communication around the world, could help to buttress our point of participatory leadership. This project, which was conceived by a Catholic priest in rural Colombia, took off with the active participation of the local populace from the very beginning. With this type of beginning, one would think that the project would be totally successful. But as we shall show shortly, its success was only partial because its initial participatory leadership model failed eventually to take into account the crucial and sustained input of local activators and servant leaders (Dagron 2001, 163–68).

The Catholic priest who founded the project, José Joaquin Salcedo Guarin, was stationed in Sutatenza. Soon after arrival, he got the idea of building a theatre in the village as an antidote to boredom and alcoholism that was plaguing the village. This would correspond to the first step in the PLCSD approach (conceptualization). The next thing he did was to open a dialogue with the parishioners as to the needs of the village. Then he introduced his idea of building a theatre. The parishioners bought the idea. This would correspond to the second step in our process (validation). The third step (legitimating) took a unique form. The priest started by projecting films in the open, and the response of the people was so positive that "in only a few weeks, a theatre was built with the contributions and active participation from the community: 1400 live chickens were donated and then sold in Bogotá to buy construction materials" (Dagron 2001, 40). The major aim of the priest was to provide literacy and education instruction to the villagers. Radio was the best means of reaching many, if not all the people. This was how Radio Sutatenza was born as part of a multimedia project that also included a printed journal.

The radio station was the first of its kind in the world, and as Dagrón rightly said, it "opened the path to thousands of participatory communication experiences" (Dagron 2001, 41). However, it failed at the final stages, that is, implementation and appropriation by the people. These would correspond to our steps 4 and 5. The people took warmly to the idea of the radio station and liked its programs. But the actual running of the project was not in their hands. This is because the leader, or the originator of the idea, was not, at the implementation stage, flying with the people like geese. Owing to that, "in reality, Radio Sutatenza did not last many years as a community-based and grassroots communication experience" (Dagron 2001, 40). The radio was soon centralized in the capital Bogotá and organizations like the World Bank and the Catholic Church, with an entrenched top-down model of communication, were involved to provide funding and operational security. With these organizations actively involved, the real participatory leadership was no longer feasible, and it was just a matter of time before the project met its downfall.

Kenya and the Green Belt Movement

In Kenya, a woman, Wangari Maathai, after planting a tree in her back garden, got the idea that tree planting could be a means to sustainable development. She thought it over, presented it to a few others, and in 1977, formed an organization known as the Green Belt Movement. Her aim was to mobilize (poor) women to plant 30 million trees to produce sustainable wood for fuel and to curb soil erosion. Getting the idea and talking it over with others and then forming an organization with them would correspond to the first and second steps in the PLCSD process (conceptualization and validation).

Once the reality check was done with a few people, and the idea was found to be sound, a movement was born: the Green Belt Movement. The idea having been validated, the next stage was to take it to the whole people for legitimating. This was done successfully, but not without difficulties. According to Maathai, who was awarded the Nobel Peace Prize for 2004, "It took me a lot of days and nights to convince people that women could improve their environment without much technology or without much financial resources." Even though Maathai was arrested, jailed, and assaulted by the then Kenyan government, her movement continued to be very successful. It went on to campaign on other related issues, like education and nutrition, that matter a lot to women. The movement successfully campaigned against a proposed building of a skyscraper in Uhuru Park, at the center of the capital city, Nairobi.

The success of the campaign against the skyscraper is proof that the people have accepted her idea. It was therefore no surprise that she went on to win a landslide victory (98% of the votes) to become a member of parliament in the Kenyan legislature in 2002, twenty-five years after she launched the Green Belt Movement. Further, she was appointed deputy environment minister in the government. The Nobel Peace Prize committee noted that she thinks globally and acts locally. They praised her for "taking a holistic approach to sustainable development that embraces democracy, human rights and women's rights in particular."[3] Although Maathai founded the Green Belt Movement, it now rightly belongs to the people. As a participatory leader, she has effectively activated a variety of stakeholders who have been able to evaluate, appropriate, and sustain her original idea. Long before she was awarded the Nobel Peace Prize, her idea had been copied and appropriated by people in other countries.

Traditional African Societies

The Kenya example makes one think of participatory leadership in traditional African societies. In precolonial Africa, communities met regularly. And because these communities were small villages, almost everybody attended those meetings with the local chiefs at the head of the meetings. Ideas, including ideas of how such communities could be developed, were brought forward by whoever had them and the people freely discussed them. The good ideas were debated, and more often than not, people reached a consensus on how to act on a particular idea for the benefit of the whole community. So, something similar to what we are proposing here was practiced in traditional Africa. Stories were the main medium used to communicate such ideas. Successful development initiatives by one community got copied and appropriated by other communities. Even though chiefs and kings ran most African nations, development initiatives did not follow the top-down model. Development ideas came from the people when they gathered at their town hall meetings. In other words, even though the political structures were more or less hierarchically fixed, with most political offices being hereditary, development ideas did not follow that pattern. Moreover, most traditional African societies were organized in groups, often according to age, gender, profession, and common interests. Original development ideas also came from members of such groups, where they were validated and then presented by that group to the whole community.

Oregon State University, Oregon

Participatory leadership for sustainable development as a tool can be applied in any setting, for any group, and for any project. To learn our way forward into the future, the PLCSD must be adopted in communities, organizations, and institutions. Oregon State University Extension Forestry uses a process that embodies the five steps of PLCSD to plan and implement educational programs. An idea for an educational program is conceived by any member of the extension forestry community (conceptualization). It is vetted with several members of the group (validating) before being brought to the whole group for approval (legitimating). Implementation only occurs if the whole group endorses the project and members of the group commit enough time and funds to carry the project to completion. Decisions are made by the group rather than by the administration. The forestry extension team annually evaluates each project and sustains it even if the membership of the group changes (appropriation). (Reichenbach and Simon-Brown 2002).

The Grameen Bank

So, among the people you have opinion leaders, servant leaders, and active citizens. As the examples from the various continents show, great ideas do not have frontiers. They can spring up from anybody who happens to have an idea. Through participatory leadership, one person's idea can be shared with and appropriated by other people, and implemented the world over. The example of the Grameen Bank, which was founded in Bangladesh by Muhammad Yunus, in similar circumstances as the Green Belt Movement, is also a case in point. The idea of micro-credit financing that empowers the poorest of the poor, which Yunus, a participatory leader, pioneered through the bank, is being copied and appropriated the world over, just like the Green Belt Movement and other similar development initiatives.

Participatory leadership for development programs or projects can take place in many ways: at the planning and decision-making level, at the implementation level, at the evaluation level, and at the outcomes sharing level (Yoon 1996, 39). But participation is a concept that is difficult to render operational. Some view it as a simple means to achieve development goals, whereas others see it as an end in itself (Melkote 1991; White 1994).

For some scholars, true participation is a basic human right, not just a way to achieve development results. It takes place when the local people are actually involved in the development process from the very beginning to the end. Within the framework of participation-as-an-end approach, many strategies were proposed, one of which was participatory action research. This strategy seeks to generate indigenous knowledge along with the dominant scientific discourse of the Western world (Awa 1989; Braun and Hoede 2000; Kibwana and et al. 2001; Melkote 2002; Tandon 1981; Turnbull 1997). The indigenous knowledge needs to be validated, shared, and maintained in order to achieve social change. But that cannot occur if the indigenous people do not have the social power. Social power is the ability to gain access, to control and influence social structures and valued social resources (Rogers 1976).

Empowerment was seen as another strategic key for participatory development. The question is: Who should have power and control over the development process? The answer to this question should be obvious: it is the community of local people. Unfortunately, no community will spontaneously gain power, control unanimously, and decide as one. This situation calls for participatory leadership.

Conclusion

Just as the modernization theory that essentially sees development as a linear process of economic growth has been discredited, the idealistic conception of participation in the development process as an alternative to modernization theory has been shown to be of very limited value in reality (Rostow 1960; Smith 2000). In many contexts, it is unrealistic to wait for a sudden collective initiative of the people. Therefore, sustainable development needs leaders. The Participatory Leadership Communication for Sustainable Development (PLCSD) process proposed in this chapter provides a realistic five-step method to achieve a full participation of the people in their development process through a growing spiral of participatory leadership communication. This process starts from an individual idea and ends with collective ownership of one or more actions. It is a working model that allows development initiators to have a clear vision about where they are and where they are going.

But the shift from "my" to "our" development initiative can occur only if there are servant leaders and active citizens who operate at the same wavelength. The leader is supposed to be humble, selfless, flexible, community minded, and nondictatorial. Such a person is not a vertical but a horizontal leader, one with the ability to articulate a clear vision, to delegate, to lead, and thus to serve. With organizational skills, he or she creates an appropriate environment conducive to the involvement of all the stakeholders. The citizens on their part are deeply interested in their community welfare. They have the ability to take on a leadership role as appropriate to the task at hand, their skill level, and the prevailing sustainable development context.

This book project started as an idea in one person's mind, and then moved from an individual's idea to a group's idea, and then to effective group action, finally resulting in a tangible product. This illustrates the power of participatory leadership for sustainable development. We thank you for reading and wish you every success in your journey toward sustainability.

Endnotes

1. This research was partly funded by the Social Sciences and Humanities Research Council of Canada (SSHRC). A different version of this paper by Charles Moumouni and Okechukwu Ukaga was published in the CD-ROM format of the proceedings of the World Congress on Communication for Development: Lessons, Challenges, and the Way Forward, organized by the Communication Initiative, the Food and Agriculture Organization of the United Nations (FAO), and the World Bank, 2007. This paper was also presented to the Annual Congress of Canadian Communication Association, Saskatoon, June 1, 2007. We acknowledge the great contribution of Professor Charles Moumouni's assistant, Uchenna Osigwe, a PhD student of philosophy at Université Laval. We thank him deeply for his comments and input.
2. The metaphor of spiral is adapted from Elizabeth Noëlle-Neumann's "spiral of silence" model (1993), but in the reverse sense. Our spiral of PLCSD model shows how an individual's idea can gain strength, credibility, and appropriation by rallying more and more support.

3. This quote and the earlier one by Wangari Maathi on being awarded the Nobel Prize can be found at http://news.bbc.co.uk/2/hi/africa/3726084.stm

References

Awa, N.J. 1989. Underutilization of women's indigenous knowledge in agriculture and rural development programs: The effect of stereotypes. In *Indigenous knowledge systems: Implications for agriculture and international development: Studies in technology and social change series*, ed. D.M. Warren, I.J. Slikkerveer, and S.O. Titilola, 186. Vol. 11. Ames: Technology and Social Change Program, Iowa State University.

Braun, A.R., and H. Hoede. 2000. Farmer participatory research in Latin America: Four cases. In *Working with farmers: The key to the adoption of forage technologies*, ed. W.W. Stur, P.M. Horme, J.B. Hacker, and P.C. Kerridge. Canberra: Australian Center for International Research. pp. 32–53.

Dagron, A.G. 2001. *Making waves*, ed. Denise Gray-Felder. New York: The Rockefeller Foundation.

Figueroa, M.E., D.L. Kincaid, M. Rani, and G. Lewis. 2002. *Communication for social change: An integrated model for measuring the process and its outcomes*. New York: The Rockefeller Foundation.

Greenleaf, R.K. 1977. *Servant leadership: A journey into the nature of legitimate power and greatness*. New York: Paulist Press.

Kibwana, O.T., et al. 2001. Clapping with two hands: Bringing local and outside knowledge for innovation in land husbandry in Tanzania and Ethiopia—A comparative case study. *Journal of Agriculture Education and Extension* 7(3).

Lazarsfeld, P.F. 1944. *Mass media and personal influence*. New York: Voice of America.

Melkote, S.R. 1991. *Communication for development in the third world: Theory and practice*. Newbury Park, CA: Sage Publications.

Melkote, S.R. 2002. Theories of developmental communication. In *Handbook of international and intercultural communication*, ed. W.B. Gudykunst and B. Mody. 2nd ed. Thousand Oaks, CA: Sage Publications. pp. 419–436.

Reichenbach, M., and V. Simon-Brown. 2002. Linking strategic thinking and project planning: The Oregon State University Extension Forestry Experience. *Journal of Extension* 40(4).

Rogers, E.M. 1976. Communication and development: The passing of the dominant paradigm. In *Communication and development: Critical perspectives*, ed. E.M. Rogers, 121–48. Beverly Hills, CA: Sage Publications.

Rogers, E.M. 1995. *Diffusion of innovations*. 4th ed. New York: Free Press.

Rostow, W.W. 1960. *The stages of economic growth, a non-communist manifesto*. Cambridge, England: University Press.

Smith, A. 2000. *An inquiry into the nature and causes of the wealth of nations*. http://www.marxists.org/reference/archive/smith-adam/works/wealth-of-nations/index.htm (accessed May 25, 2008).

Tandon, R. 1981. Participatory research in the empowerment of people. *An International Journal of Adult Education* 14(3):20–29.

Turnbull, D. 1997. Reframing science and other local knowledge traditions. *Futures* 29:551–62.

Ukaga, O. 2005. Preparing Africa for sustainable development in the new millennium. In *Sustainable development in Africa: A multifaceted challenge*, ed. O. Ukaga and O.G. Afoaku, 239. Trenton, NJ: Africa World Press.

Ukaga, O., and C. Maser. 2004. *Evaluating sustainable development: Giving people a voice in their destiny*. Sterling, VA: Stylus Pub.

White, S.A. 1994. Introduction: The concept of participation. In *Participatory communication: Working for change and development*, ed. S.A. White, K.S. Nair, and J.R. Ascroft, 15–32. Thousand Oaks, CA: Sage.

World Commission on Environment and Development (WCED). 1987. *Our common future*. Oxford: Oxford University Press.

Yoon, C.S. 1996. La Comunication Participative Pour le Developpement. In *La Communication Participative pur le Developpemnet: Un Agenda Ouest-Africain*, ed. G. Bessette and C. V. Rajasunderam, 39–53. Ottawa: Centre de Recherche pour le Developpement International, CRDI.

Index

N

O

P

Q

R

S

T

www.ingramcontent.com/pod-product-compliance
Ingram Content Group UK Ltd.
Pitfield, Milton Keynes, MK11 3LW, UK
UKHW022000180926
475701UK00003B/121
* 9 7 8 0 3 6 7 3 8 4 5 4 8 *